Heat and Mass Transfer in Building Services Design

Other titles from E & FN Spon

Facilities Management: Theory and practice
Edited by Keith Alexander

Ventilation of Buildings
Hazim B. Awbi

Air Conditioning: A practical introduction
2nd edition
David V. Chadderton

Building Services Engineering
2nd edition
David V. Chadderton

Building Services Engineering Spreadsheets
David V. Chadderton

Naturally Ventilated Buildings
Derek Clements-Croome

Illustrated Encyclopedia of Building Services
David Kut

Spon's Mechanical and Electrical Services Price Book 1988
Davis Langdon & Everest

Building Energy Management Systems
Geoff Levermore

Energy Management and Operating Costs in Buildings
Keith J. Moss

Heating and Water Services Design in Buildings
Keith J. Moss

International Dictionary of Heating, Ventilation and Air Conditioning
2nd edition
REHVA

Sick Building Syndrome
Jack Rostron

Site Management of Building Services Contractors
Jim Wild

For more information about these and other titles please contact:
The Marketing Department, E & FN Spon, 11 New Fetter Lane, London, EC4P 4EE. Tel: 0171 583 9855

Heat and Mass Transfer in Building Services Design

Keith J. Moss I. ENG., ACIBSE
Visiting lecturer in Building Services Engineering to
The City of Bath College and the University of Bath

E & FN SPON
An Imprint of Routledge
London and New York

First published 1998
by E & FN Spon, an imprint of Routledge
11 New Fetter Lane, London EC4P 4EE

Simultaneously published in the USA and Canada
by Routledge
29 West 35th Street, New York, NY 10001

© 1998 Keith J. Moss

Typeset in 10½/12pt Sabon by
Pure Tech India Ltd, Pondicherry
Printed in Great Britain by
The Alden Press, Oxford

All rights reserved. No part of this book may be reprinted or reproduced or utilized in any form or by any electronic, mechanical, or other means, now known or hereafter invented, including photocopying and recording, or in any information storage or retrieval system, without permission in writing from the publishers.

British Library Cataloguing in Publication Data
A catalogue record for this book is available from the British Library

ISBN 0 419 22650 8

Contents

Preface	ix
Acknowledgements	x
Introduction	xi

1 Thermal comfort and assessment — 1
Nomenclature — 1
1.1 Introduction — 1
1.2 Heat energy and temperature — 2
1.3 Thermometry — 3
1.4 Types of thermometer — 3
1.5 Heat loss from the human body — 4
1.6 Physiological responses — 9
1.7 Thermal assessment — 9
1.8 Thermal comfort criteria — 14
1.9 Temperature profiles — 22
1.10 Chapter closure — 23

2 Heat conduction — 24
Nomenclature — 24
2.1 Introduction — 25
2.2 Heat conduction at right angles to the surface — 25
2.3 Surface conductance — 29
2.4 Heat conduction in ground floors — 33
2.5 Heat conduction in suspended ground floors — 35
2.6 Thermal bridging and non-standard U values — 37
2.7 Non-standard U values, multi-webbed bridges — 39
2.8 Radial conductive heat flow — 42
2.9 Chapter closure — 48

3 Heat convection — 50
Nomenclature — 50
3.1 Introduction — 50
3.2 Rational formulae for free and forced convection — 53
3.3 Temperature definitions — 54
3.4 Convective heat output from a panel radiator — 57
3.5 Heat output from a pipe coil freely suspended — 58
3.6 Heat transfer from a tube in a condensing secondary fluid — 60
3.7 Cooling flux from a chilled ceiling — 62

	3.8	Heat flux off a floor surface from an embedded pipe coil	63
	3.9	Chapter closure	66
4	**Heat radiation**	**67**
	Nomenclature	67
	4.1	Introduction	67
	4.2	Surface characteristics	68
	4.3	The greenhouse effect	71
	4.4	Spectral wave forms	71
	4.5	Monochromatic heat radiation	72
	4.6	Laws of black body radiation	73
	4.7	Laws of grey body radiation	74
	4.8	Radiation exchange between a grey body and a grey enclosure	75
	4.9	Heat transfer coefficients for black and grey body radiation	76
	4.10	Heat radiation flux I	77
	4.11	Problem solving	78
	4.12	Asymmetric heat radiation	88
	4.13	Solar irradiation and the solar constant	89
	4.14	Solar collectors	91
	4.15	Chapter closure	92
5	**Measurement of fluid flow**	**94**
	Nomenclature	94
	5.1	Introduction	94
	5.2	Flow characteristics	94
	5.3	Conservation of energy in a moving fluid	96
	5.4	Measurement of gauge pressure with an uncalibrated manometer	97
	5.5	Measurement of pressure difference with an uncalibrated differential manometer	98
	5.6	Measurement of flow rate using a venturi meter and orifice plate	100
	5.7	Measurement of air flow using a pitot static tube	106
	5.8	Chapter closure	108
6	**Characteristics of laminar and turbulent flow**	**109**
	Nomenclature	109
	6.1	Introduction	109
	6.2	Laminar flow	110
	6.3	Turbulent flow	112
	6.4	Boundary layer theory	114
	6.5	Characteristics of the straight pipe or duct	118
	6.6	Determination of the frictional coefficient in turbulent flow	119

	6.7	Solving problems	120
	6.8	Chapter closure	127
7	**Flow of fluids in pipes, ducts and channels**		**129**
	Nomenclature	129
	7.1	Introduction	130
	7.2	Solutions to problems in frictionless flow	130
	7.3	Frictional flow in flooded pipes and ducts	136
	7.4	Semi-graphical solutions to frictional flow in pipes and ducts	148
	7.5	Gravitational flow in flooded pipes	150
	7.6	Gravitational flow in partially flooded pipes and channels	156
	7.7	Alternative rational formulae for partial flow	162
	7.8	Chapter closure	164
8	**Natural ventilation in buildings**		**165**
	Nomenclature	165
	8.1	Introduction	166
	8.2	Aerodynamics around a building	167
	8.3	Effects on cross ventilation from the wind	170
	8.4	Stack effect	173
	8.5	Natural ventilation to internal spaces with openings in one wall only	176
	8.6	Ventilation for cooling purposes	178
	8.7	Fan assisted ventilation	183
	8.8	Chapter closure	183
9	**Regimes of fluid flow in heat exchangers**		**184**
	Nomenclature	184
	9.1	Introduction	185
	9.2	Parallel flow and counterflow heat exchangers	185
	9.3	Heat transfer equations	188
	9.4	Heat exchanger performance	195
	9.5	Cross flow	200
	9.6	Further examples	202
	9.7	Chapter closure	206
10	**Verifying the form of an equation by dimensional analysis**		**207**
	Nomenclature	207
	10.1	Introduction	207
	10.2	Dimensions in use	208
	10.3	Chapter closure	211
11	**Solving problems by dimensional analysis**		**213**
	Nomenclature	213
	11.1	Introduction	213

11.2	Establishing the form of an equation	213
11.3	Dimensional analysis in experimental work	215
11.4	Examples in dimensional analysis	217
11.5	Chapter closure	228

Bibliography	229
Index	231

Preface

This is the last book in the trilogy that began with *Heating and Water Services Design in Buildings* and continued with *Energy Management and Operating Costs in Buildings*. In this book I have endeavoured to provide text, problems and solutions that relate the subjects of heat and mass transfer to the discipline of building services engineering.

While there is currently a shift towards making people multi-skilled, this does not infer that programmes of learning can become generic. Indeed it is increasingly the case that courses are now more specific to the needs of the individual at the workplace. At least one national awarding authority has emphasized for some time that their programmes of study should be related to workplace activities.

This provides a particular challenge to lecturers, teachers and authors who deliver ancillary subjects like mathematics, engineering science and thermofluids. Last year was dedicated to YES, the year of engineering success and in his presidential address Jerome O'Hea of the Chartered Institution of Building Services Engineers said that in essence we need to convey the good news that a career in building services engineering is a rich and rewarding one.

I strongly feel that as a practitioner, lecturer and author I have a commitment also to ensuring that the learning experience is rich and rewarding, and this in my view is primarily achieved through making ancillary subjects which underpin the primary subjects of a course of study, current and relevant.

Acknowledgements

I acknowledge with thanks the permission granted by the Chartered Institution of Building Services Engineers to reproduce some data from the *CIBSE Guide* (1986).

The material used in this book has been prepared and amended over many years. If copyright permission has been overlooked, the publishers will be pleased to take up the matter on request.

I have to acknowledge the patience of my teachers. In particular my thanks are due to Alec Griffiths who taught me the basic rules of mathematics and then showed me how mathematics can be used as a tool for solving practical problems. Another person whose name escapes me was an Australian lecturer at what was then the Southbank Polytechnic, who in 1960, with his warmth of personality and extraordinary ability to teach, inspired the whole group of us mathematically ignorant students who by the end of the course were sufficiently competent, and motivated, to successfully pass the examination.

Introduction

This book is intended to provide, within the limits of its title, the underpinning knowledge for the technology subjects of space heating, water services, ventilation and air conditioning.

The reader will find that it is necessary to participate and respond to the narrative which has been written for those with a good grounding in mathematics who have an interest, vested or otherwise, in the technology subjects.

With the explosion of IT in the form of dedicated software for design purposes it is very easy not to give sufficient attention to fundamental theory and even design calculations. A balance in the process of course delivery has somehow to be struck between the acquisition of underpinning knowledge and developing the skills required to use dedicated software and computer aided design systems.

One way to achieve a balance for the student is to ensure that the support subjects like heat and mass transfer are dedicated to relating fundamental principles to practical design applications. This will help to secure an interest at least in an important part of the learning process.

If after reading and participating in parts or all of this book, it has provided a learning experience and the reader has been enthused by even a little, my efforts will have been rewarded.

Thermal comfort and assessment 1

Nomenclature

A	area (m²)
ASHRAE	American Society of Heating Refrigeration & Air Conditioning Engineers
Clo	unit of thermal resistance of clothing
dt	temperature difference (K)
eh_r	heat transfer coefficient for radiation (W/m²K)
emf	electromotive force (V)
h	height (m)
h_c	heat transfer coefficient for convection (W/m²K)
m	mass (kg)
Met	metabolic rate
mwet	mean weighted enclosure temperature (°C)
P	pressure (Pa)
PMV	predicted mean vote
PPD	predicted percentage dissatisfied
Q	rate of heat loss (W)
Q_c	rate of heat loss/gain by convection (W)
Q_{cd}	rate of heat loss/gain by conduction (W)
Q_e	rate of heat loss by evaporation (W)
Q_r	rate of heat loss/gain by radiation (W)
t	temperature (°C)
t_a	air temperature (°C)
t_{am}	ambient temperature (°C)
t_c	comfort temperature (°C)
t_e	environmental temperature (°C)
t_g	globe temperature (°C)
t_r	mean radiant temperature (°C)
u	mean air velocity (m/s)
V	volume (m³)

1.1 Introduction

This first chapter introduces you to temperature, the variations of which provides the motive force in heat transfer, and heat energy, the flow and transport of which in air, water and steam is the essence of much of heating, ventilating and air conditioning design. Its main focus, however, is on the topic of thermal comfort and the assessment of indoor climates, in which people live and work, to establish levels of comfort. ASHRAE define thermal comfort as 'that condition of mind in which satisfaction is expressed with the thermal environment'. The accurate assessment of building heat losses and gains, the type of comfort systems selected and the regimes of control of the comfort systems are all directed toward achieving this definition.

The effect that the amount of clothing which is worn has on different levels of activity also impinges on the comfort of the individual. Because thermal comfort is also a subjective assessment, a minority of individuals may feel uncomfortable even in thermal environments which are well regulated.

1.2 Heat energy and temperature

A definition of energy is the capacity a substance possesses which can result in the performance of work. It is a property of the substance. Heat on the other hand is energy in transition. Heat is one form of energy and can be expressed, for example, as a specific heat capacity in kJ/kgK. In this form it is expressing the potential of a substance for storing heat which it has absorbed from its surroundings. It can also express the potential for the intensity of heat transfer from the substance to its surroundings.

Up until the end of the 18th century heat energy, known as 'caloric', was considered as a fluid which could be made to flow for the purposes of space heating among other things, or it flowed of its own volition as a result of friction which was generated as a result of a process or work done, such as boring out a cannon. The idea of heat being a form of energy rather than a fluid was developed by an American named Benjamin Thompson, subsequently known as Count Rumford, during the process of boring out cannons for his arsenal as war minister of Bavaria. His conclusions were that the amount of heat liberated depended upon the work done against friction by the boring device.

A partial definition of heat energy is therefore the interaction between two substances which occurs by virtue of their temperature difference when they communicate. However, heat energy does not always initiate a rise in temperature as in the cases of the latent heat of vaporization and condensation which occur when substances change in state. This is a qualification of the definition.

Heat is a transient commodity like work; it exists during communication only although like work its effect may be permanent. The primary need for burning fuel oil might be the generation of heat energy and the permanent results of the process are the products of combustion. The combustion products cannot return to fuel oil. The transient result is the generation of heat.

A definition of temperature is a scaled measurement of relative hotness and coldness sensations. It can be described as an intensity

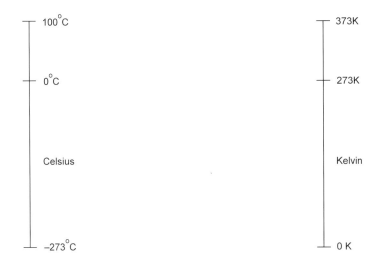

Figure 1.1 Scales of temperature.

of hotness or coldness. Kelvin found that absolute coldness is reached when the agitation of the molecules and atoms of a substance ceases at −273.15°C (0 K).

Scales of temperature have been advanced by various authorities. The scales now commonly in use are the Celsius scale and the Kelvin scale. Figure 1.1 shows these scales from absolute zero to the upper fixed point.

1.3 Thermometry

In the 17th century it was proposed that two fixed points should be used to determine a temperature scale.

1. The lower fixed point was taken as melting ice at atmospheric pressure; the ice being distilled water.
   The ice point is the temperature at which ice and water can exist in equilibrium.
2. The upper fixed point was taken as steam generated from distilled water when boiling at atmospheric pressure.

Since the temperature tends to vary depending upon geographical location the steam point is the temperature of boiling water and steam at atmospheric pressure on latitude 45°. Table 1.1 lists the fixed points of the International Temperature Scale at standard atmospheric pressure (101 325 Pa).

1.4 Types of thermometer

There are six main types of thermometer:

- constant volume gas thermometer
- resistance thermometers

4   Thermal comfort and assessment

Table 1.1 Fixed points of the International Temperature Scale

Fixed points	Temperature (°C)
Boiling point of liquid oxygen	−182
Ice point	0
Steam point	100
Boiling point of sulphur	444.6
Freezing point of silver	960.8
Freezing point of gold	1063.0

- thermocouples
- liquid thermometers
- bimetallic thermometer
- pyrometers.

The constant volume gas thermometer was selected in 1887 as the standard: it did not give a pointer reading, however. For a perfect gas Boyle's law states that $P \propto 1/V$ at constant temperature and the Kelvin scale (Figure 1.1) agrees exactly with the scale of a perfect gas thermometer.

The resistance thermometer consists of platinum wire wound on to two strips of mica and the coil attached to leads connected in turn to a Wheatstone bridge.

Thermocouples have the measuring element as the junction between two dissimilar metal wires. The emf generated at the junction results from its temperature and this is measured accurately on a potentiometer or approximately using a galvanometer.

The liquid thermometer relies on the expansion of a liquid in a glass or steel tube in response to a rise in temperature.

The bimetallic thermometer is associated with dial instruments in which two dissimilar metal strips soldered together in a coil expand differentially on a rise in temperature thus moving a pointer round the dial.

Pyrometers, of which there are four types, tend to be used for measuring very low temperatures and temperatures up to 1400°C. The four types are: resistance, thermoelectric, radiation and optical.

1.5 Heat loss from the human body

The core temperature of the human body is taken as 37.2°C. This implies that humans as with all mammals must generate heat in order to maintain body temperature. About 80% of food intake is required to maintain body temperature.

The metabolic rate refers to the rate at which energy is released from food into the body cells. It is affected by a person's size, body fat, sex, age, hormones and level of activity. Table 1.2 lists the approximate total heat output for different levels of activity. The Met is equal to the

Table 1.2 Heat emission from the human body (adult male, body surface area 2 m²)

Degree of activity	Application Typical	Total	Sensible (s) and latent (l) heat emissions/W at the stated dry-bulb temperatures/°C									
			15		20		22		24		26	
			(s)	(l)	(s)	(l)	(s)	(l)	(s)	(l)	(s)	(l)
Seated at rest	Theatre, Hotel lounge	115	100	15	90	25	80	35	75	40	65	50
Light work	Office, Restaurant*	140	110	30	100	40	90	50	80	60	70	70
Walking slowly	Store, Bank	160	120	40	110	50	100	60	85	75	75	85
Light bench work	Factory	235	150	85	130	105	115	120	100	135	80	155
Medium work	Factory, Dance hall	265	160	105	140	125	125	140	105	160	90	175
Heavy work	Factory	440	200	220	190	250	165	275	135	305	105	335

*For restaurants serving hot meals, add 10 W sensible and 10 W latent for food.
Reproduced from the *CIBSE Guide* (1986) by permission of the Chartered Institution of Building Services Engineers.

Table 1.3 Metabolic rate for different levels of activity

Activity	Metabolic Rate	
	Met	W/m²
Lying down	0.8	45
Seated quietly	1.0	58
Sedentary work – seated at work	1.2	70
Light activity – bodily movement on foot	1.6	93
Medium activity – bodily movement including carrying	2.0	117
High activity – substantial physical work	3.0	175

metabolic rate for a seated adult at rest and is equivalent to 58 W/m² of body surface. Table 1.3 lists the metabolic rate in units of Met and W/m² for different levels of activity.

In order to estimate the heat loss from a person's body its surface area is required. A close approximation of surface area A is obtained from a knowledge of the person's height h and mass m using an empirical formula attributed to Dubois where:

$$A = (m^{0.425} \times h^{0.725} \times 0.2024)\,\text{m}^2$$

Thus for a person of mass 70 kg and height 1.8 m

$$A = 6.1 \times 1.53 \times 0.2024 = 1.89\,\text{m}^2$$

A figure of 2 m² is frequently used for the surface area of a clothed adult.

Discomfort will occur if the body temperature varies much from the core temperature of 37.2°C. A loss of about 2.5 K for an extended period may induce a state of hypothermia. In the old and the young particularly this is a serious condition leading ultimately to death. A state of hypothermia means that the body cannot restore itself to its

6 Thermal comfort and assessment

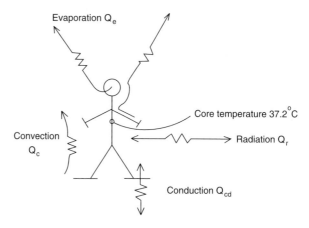

Figure 1.2 Latent heat loss and sensible heat loss/gain from the body.

normal temperature without the aid of active heating. Being put into a pre-warmed bed which is then maintained at a constant temperature for example. The intake of high calorie hot food will also be necessary to aid the recovery process. Wrapping the sufferer in warm and reflective clothing may be a temporary measure but not one which will restore the patient's body temperature.

If body temperature rises by 4 K for an extended period a state of hyperpyrexia may be induced. In this condition the body is unable to liberate its heat sufficiently to its surroundings and unconsciousness may result. Hyperpyrexia can be induced internally through a fever or externally from high ambient temperature, high humidity and the effects of solar radiation incident upon the head in particular.

A person is likely to be comfortable in the general sense of the word when the heat generated by the body to maintain a core temperature of 37.2°C is equal to the heat lost to its surroundings.

The heat generated Q can be expressed as:

$$Q = \pm Q_c \pm Q_{cd} \pm Q_r + Q_e. \quad \text{See Figure 1.2.}$$

Heat conduction Q_{cd} takes place at points of physical contact. It constitutes a small proportion of the total and is usually ignored, thus:

$$Q = \pm Q_c \pm Q_r + Q_e$$

Sensible heat gain or loss therefore includes heat convection and heat radiation. Latent heat loss from the body occurs at all times and results in evaporative cooling. It includes insensitive perspiration from the skin surface, moisture evaporation from the process of breathing, and sweating.

Latent heat loss from the body is in three forms, namely:

- passive moisture loss from the skin which depends upon the vapour pressure of the surrounding air.

- moisture loss from the lungs which depends upon ambient vapour pressure and breathing rate which in turn depends upon the degree of activity and thus the metabolic rate.
- active sweating commences when sensible heat loss plus insensible perspiration fall below the body's rate of heat production.

Sweat is secreted by the eccrine glands which lie deep in the skin tissue. It consists of 99% water and 1% sodium chloride. The eccrine glands are activated by two control mechanisms:

> stimulus – peripheral receptors or sympathetic nerves, and
> thermoregulation – the hypothalamus which is located in the brain and which responds to its own temperature variations.

The eccrine glands can be activated by heat energy resulting from physical work or the local climate or by physiological stimuli especially those located in the palms of the hands, the soles of the feet, the face and the chest.

For men working in the heat the sweat rate can reach 1 litre/h. Hard work in a very hot environment may increase this rate to 2.5 l/h. This rate cannot be sustained for more than 30 min. Thus the bodily heat loss is restricted as the period of hard labour continues and core temperature is elevated.

Refer to Table 1.2 for rates of body heat loss for different levels of activity. You will notice in this table that as the air temperature rises from 15°C to 26°C the proportions of latent to sensible heat change but the total body heat loss for each level of activity does not change with air temperature rise. Thus for light work the total heat loss is 140 W and the ratio of latent to sensible heat loss at an air temperature of 15°C is $30/110 = 0.27$ whereas at an air temperature of 26°C the ratio is $70/70 = 1.0$. This represents a 3.7-fold increase in insensitive perspiration and sweat assuming moisture evaporation from breathing is unchanged. You will be able to corroborate this evidence from personal experience of working or even sitting in surroundings of relatively high air temperature.

Figure 1.3 is a flow chart showing sensible and latent heat flows from the human body to the surrounding climate. Note the thermal criteria which trigger each mode of heat transfer.

It was stated earlier that a heat balance can be drawn such that the heat generated to maintain a core temperature of 37.2°C is equal to the bodily heat loss to its surroundings.

In thermally comfortable surroundings when mean radiant, air and comfort temperature are say 20°C, a person doing light work absorbs approximately 24 l of oxygen per hour. For each litre of oxygen consumed, 21 kJ of heat are produced.

Thus heat generated by the person $= 24 \times 21 \times 1000/3600 = 140$ W

8 Thermal comfort and assessment

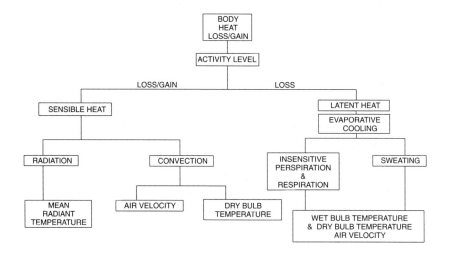

Figure 1.3 Heat energy flows between the human body and the surrounding climate.

Heat loss to the surroundings

$$Q = \pm Q_c \pm Q_r \pm Q_e$$

if the average temperature of exposed skin and clothing is 26°C and the surface area of the clothed body is 2m².

Sensible heat loss

$$Q_c = h_c.A.dt$$
$$Q_c = 3 \times 2 \times (26 - 20) = 36\,\text{W}$$
$$Q_r = eh_r.A.dt$$
$$Q_r = 5.7 \times 2 \times (26 - 20) = 68.4\,\text{W}$$

Latent heat loss
If the person loses 0.05 l/h and the latent heat of vaporization is taken as 2500 kJ/kg

$$Q_e = (0.05/3600) \times 2500 \times 1000 = 34.7\,\text{W}$$

Thus body heat loss

$$Q = 36 + 68.4 + 34.7$$
$$Q = 139\,\text{W}$$

and the heat generated by the body, calculated as 140 W, shows that the heat balance is maintained.

You should refer to Chapters 3 and 4 for the heat transfer coefficients h_c and eh_r for convection and radiation.

1.6 Physiological responses

Circulatory regulation of the blood flow is the initial response to thermal stress, and in the subcutaneous layer which connects the skin to the surface muscles is known as vasomotor regulation. Regulation of the blood flow is achieved by vasodilation and vasoconstriction of the blood vessels. The vasomotor centre is located in part of the brain known as the medulla oblongata.

The subcutaneous tissue has a high fat content and thus a high resistance to heat flow through the subcutaneous layer. Thus vasodilation within the subcutaneous layer induces large quantities of blood from the core through the subcutaneous tissue to the skin giving rise to high heat energy rejection. Vasoconstriction within the subcutaneous tissue induces low blood flow from the body core to the skin. These regulatory effects on the flow of blood vary the resistance to heat flow through the subcutaneous tissue. Thus the thermal resistance of the subcutaneous tissue which controls the heat flow at the skin surface is variable and responds to the degree of activity and ambient temperature in the manner described.

Low ambient temperature induces vasoconstriction in the subcutaneous tissue and hence lowers skin temperature for a person doing sedentary work. This inhibits excessive body heat loss to preserve core temperature. The sensation is the cooling of the extremities – fingers, nose, ears and toes.

Vasodilation is accompanied by an increase in heart rate, an increase in blood flow to the skin resulting in increased body heat loss and a reduced blood flow to the organs and is induced by a high level of physical activity.

1.7 Thermal assessment

Human thermal comfort is a subjective condition which is witnessed by most of us fairly regularly: witness the varying amounts of clothing worn by different people in the same room. It is generally accepted, however, that there are four criteria which have a direct influence on human comfort:

- dry bulb temperature
- wet bulb temperature
- mean radiant temperature
- air velocity.

Each of these criteria can be varied within limits, and still maintain comfort level, to compensate for one of them having a value outside the comfort range.

Many proposals for a thermal index which accounts for some or all of these criteria have been advanced in the last 100 years. The thermal indices in current use are dry resultant or comfort temperature and environmental temperature although the latter is not now used for the purposes of measurement.

Air temperature is that measured by a mercury in glass thermometer shielded from direct heat radiation and suspended in air. The sensing bulb is small and as the mercury is reflective anyway, heat radiation incident on the bulb surface is insignificant, allowing the bulb to register local air temperature.

Wet bulb temperature is obtained by placing a muslin sock over the sensing bulb of mercury in a glass thermometer and saturating it by placing the end of the sock in a container of distilled water. If the air local to the sensing bulb is dry it has a low relative humidity and will evaporate the moisture in the sock. The rate of evaporation produces a cooling effect which will depress the mercury in the thermometer thus giving the wet bulb reading. The rate of evaporation on the sock of a wet bulb thermometer is proportional to the level of moisture in the local air, and a reading equal to the local dry bulb temperature implies that evaporation has ceased because the local air is saturated and relative humidity is therefore 100%.

The humidity range to ensure a satisfactory level of comfort is between 40% and 70%. The hand-held whirling sling psychrometer is still a popular instrument used for measuring dry bulb and wet bulb temperatures.

Mean radiant temperature can be approximately evaluated from the mean weighted enclosure temperature, mwet:

$$\text{mwet} = (A_w t_w + A_f t_f + A_r t_r + A_g t_g)/(A_w + A_f + A_r + A_g)$$

thus

$$\text{mwet} = \Sigma(A t_s)/\Sigma A$$

The suffixes refer to the surface temperature of the enclosing walls, floor, roof and glazing.

You can see that it is the mean area weighted temperature of all the surfaces forming the enclosure. It is approximate because it is only most nearly a true mean radiant temperature if the point of measurement is in the centre of the space.

Mean radiant temperature can be measured with the aid of a globe thermometer (Figure 1.4). The matt-finished sensing surface is greatly enlarged and is therefore more sensitive to absorb heat radiation than to sense the temperature of the air local to it. As air velocity increases, however, this instrument becomes less effective at measuring mean radiant temperature. See Example 1.1. The formula for globe temperature is:

globe temperature

$$t_g = (t_r + 2.35 t_a(u)^{0.5})/(1 + 2.35(u)^{0.5})$$

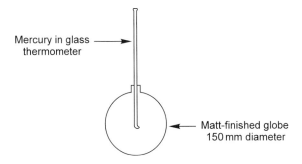

Figure 1.4 The globe thermometer.

from which
$$t_r = t_g(1 + 2.35(u)^{0.5}) - (2.35t_a(u)^{0.5})$$
when air velocity
$$u = 0.1\,\text{m/s}, t_g = 0.57t_r + 0.43t_a$$
and
$$t_r = 1.75t_g - 0.75t_a$$
You should confirm that you agree with these three equations which originate from the equation for globe temperature t_g.

Comfort temperature is measured by a similar instrument to that for globe temperature except that the sensing surface is a 100 mm sphere (Figure 1.4). It is also known as dry resultant temperature and was introduced by a Frenchman in 1931. The formula for comfort temperature is:
$$t_c = (t_r + t_a(10u)^{0.5})/(1 + (10u)^{0.5})$$
Another formula for t_c is given in the *CIBSE Guide* section A, 1970 edition, as:
$$t_c = (t_r + 3.17t_a(u)^{0.5})/(1 + 3.17(u)^{0.5})$$
This formula is in the same format as that for globe temperature. Either formula will give the same results. When air velocity $u = 0.1\,\text{m/s}, t_c = 0.5t_r + 0.5t_a$

Note: the above formulae for globe temperature and comfort temperature are empirical and therefore subject to some error.

Environmental temperature is not easily measured by an instrument and this is the main reason why it is not in common use now. For a cubical room in which local air velocity is around 0.1m/s
$$t_e = 0.667t_r + 0.333t_a$$

Low air speed is measured using a Kata thermometer. The time taken for warmed fluid to cool and contract down a glass stem between two fixed points is noted and use made of a nomogram provided by the manufacturer to convert the time to an air speed.

Note: electronic instruments for measuring t_a, t_r, t_w, t_c and u are available.

You will see that comfort temperature and environmental temperature account for three of the four factors which influence thermal comfort. These are air temperature, mean radiant temperature and air velocity. Wet bulb temperature and hence relative humidity and vapour pressure is not accounted for and cannot be measured by a comfort temperature instrument. However, comfort temperature is currently the accepted thermal index in space heating design. A separate instrument is required to register relative humidity. You will notice from the formula that at an air speed of 0.1 m/s which is an acceptable value in a room not subject to forced air movement, comfort temperature represents the sum of 50% air temperature and 50% mean radiant temperature. The formula for environmental temperature on the other hand is weighted towards mean radiant temperature. For higher air speeds the equations for t_c and t_g change by varying the proportions of air and mean radiant temperature. Consider the following example.

Example 1.1
Determine the equations for comfort temperature and mean radiant temperature when air velocity is found to be 0.4 m/s.

Solution
It was shown earlier that:

$$t_g = (t_r + 2.35 t_a (u)^{0.5})/(1 + 2.35(u)^{0.5})$$

substituting $u = 0.4\,\text{m/s}$

$$t_g = t_r + 2.35 t_a (0.4)^{0.5})/(1 + 2.35(0.4)^{0.5})$$

$$t_g = (t_r + 1.4863 t_a)/2.4863$$

$$t_g = (t_r/2.4863) + (1.4863 t_a/2.4863)$$

from which

$$t_g = 0.4 t_r + 0.6 t_a$$

$$t_r = (t_g - 0.6 t_a)/0.4$$

from which

$$t_r = 2.5 t_g - 1.5 t_a$$

Now, from earlier in this section it was shown that:

$$t_c = (t_r + t_a (10u)^{0.5})/(1 + (10u)^{0.5})$$

substituting u = 0.4 m/s
$$t_c = (t_r + 2t_a)/3$$
from which
$$t_c = 0.33 t_r + 0.67 t_a$$

It is now appropriate to analyse the effects that air temperature and globe temperature have upon mean radiant temperature and comfort temperature at different air velocities. Consider the following case study.

Case study 1.1

A heated room is used for sedentary occupation.
a) Evaluate comfort and mean radiant temperature for the room in which the measured globe and air temperatures are 17°C and 21°C respectively for air velocities of 0.1 m/s and 0.4 m/s.
b) Evaluate comfort and mean radiant temperature for the room in which the measured globe and air temperatures are 21°C and 17°C respectively for air velocities of 0.1 m/s and 0.4 m/s.
c) Summarize and draw conclusions from the results.

SOLUTION

a) When $t_g = 17°C$, $t_a = 21°C$ and $u = 0.1$ m/s
$$t_r = 1.75 t_g - 0.75 t_a = 1.75 \times 17 + 0.75 \times 21 = 14°C$$
$$t_c = 0.5 t_r + 0.5 t_a = 0.5 \times 14 + 0.5 \times 21 = 17.5°C$$
when $u = 0.4$ m/s
$$t_r = 2.5 t_g - 1.5 t_a = 2.5 \times 17 - 1.5 \times 21 = 11°C$$
$$t_c = 0.33 t_r + 0.67 t_a = 0.33 \times 11 + 0.67 \times 21 = 17.67°C$$
b) When $t_g = 21°C$, $t_a = 17°C$ and $u = 0.1$ m/s
$$t_r = 1.75 \times 21 - 0.75 \times 17 = 24°C$$
$$t_c = 0.5 \times 24 + 0.5 \times 17 = 20.5°C$$
when $u = 0.4$ m/s
$$t_r = 2.5 \times 21 - 1.5 \times 17 = 27°C$$
$$t_c = 0.33 \times 27 + 0.67 \times 17 = 20.3°C$$
c) The analysis is summarized in Table 1.4

Table 1.4 Summary of case study 1.1

Measured u (m/s)	Measured °C		Calculated °C		Case	Heat flow paths				$(t_r - t_a)$ temp. diff. (K)
	t_g	t_r	t_a	t_c						
0.1	17	14	21	17.5	1	t_a	t_c	t_g	t_r	−7
0.4	17	11	21	17.67	2	t_a	t_c	t_g	t_r	−10
0.1	21	24	17	20.5	3	t_r	t_g	t_c	t_a	7
0.4	21	27	17	20.3	4	t_r	t_g	t_c	t_a	10

CONCLUSIONS FROM THE SUMMARY OF CASE STUDY 1.1

There are two comfort zones which can be applied here.

- For mainly sedentary occupations comfort temperature t_c should fall between 19 and 23°C.
  Cases 1 and 2 therefore fall outside this comfort zone and may not be conducive to thermal comfort.
- The difference in temperature between mean radiant and air temperature should be within an envelope of +8 K or −5 K for mainly sedentary occupations.
  Cases 1, 2 and 4 therefore fall outside this comfort zone and may not provide a satisfactory level of thermal comfort.
  Case 3 thus appears to be the only right solution here for sedentary occupation.
  Current standards of thermal insulation recommended in the Building Regulations along with limiting the infiltration rate of outdoor air will mitigate in favour of maintaining the difference between mean radiant temperature and air temperature within the +8 K, −5 K envelope.
  This matter is discussed in detail in *Heating and Water Services Design in Buildings*.
- The ranking of the thermal indices in the heat flow paths is dependent upon the type of space heating. Cases 3 and 4 indicate a mainly radiant system of space heaters; cases 1 and 2 indicate a mainly convective heating system.
- The most acceptable level of comfort for sedentary occupations is achieved when the difference between mean radiant and air temperature is within a +8 K, −5 K envelope and the comfort temperature zone of between 19°C and 23°C, with relative humidity between 40% and 70%.
- If electronic temperature measurement equipment is used you can see the importance of establishing which thermal index the instrument is actually measuring.

1.8 Thermal comfort criteria

Professor O. Fanger has spent much time both in Denmark and the USA researching how thermal comfort can be assessed for a variety of occupations.

The metabolic rate can vary from 30 W/m² to 500 W/m² depending upon the level of activity. One Met is equal to the metabolic rate for a seated person and is equivalent to 58 W/m².

The thermal resistance of clothing is measured in units of 'Clo' such that:

one Clo unit = 0.155 m²K/W.

Table 1.5 gives typical insulation levels for some clothing ensembles.

Table 1.5 Clothing insulation levels – the relationship between the Clo unit and thermal resistance R

Clothing combination	Insulation level	
	Clo	R (m²K/W)
Naked	0	0
Shorts/bikini	0.1	0.016
Light summer clothing	0.5	0.078
Indoor winter clothing	1.0	0.155
Heavy suit	1.5	0.233
Polar weather suit	3–4	0.465–0.62

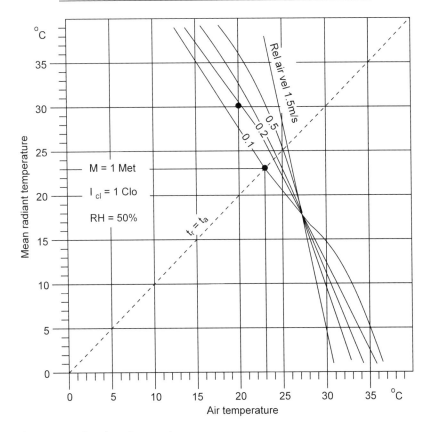

Figure 1.5 Comfort diagram 1.

16 Thermal comfort and assessment

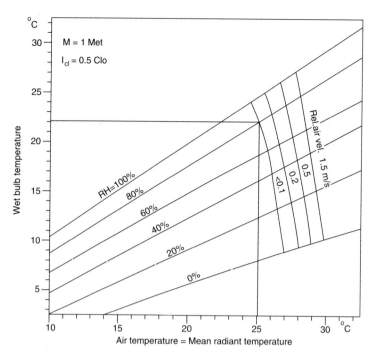

Figure 1.6 Comfort diagram 2.

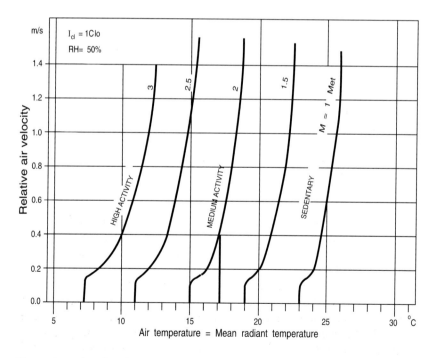

Figure 1.7 Comfort diagram 3.

In his research in thermal assessment Professor Fanger uses the term ambient temperature (which is often used as a term for outdoor temperature) to refer to indoor conditions when air and mean radiant temperature are the same in value.

The comfort diagrams which he produced can be used by HVAC engineers as well as those practising occupational hygiene and by health and safety officers.

Three of the diagrams are shown in Figures 1.5, 1.6 and 1.7 and the following examples will demonstrate their use.

Example 1.2
The staff occupying an office are clothed at 1 Clo and are undertaking sedentary activities. Determine the required ambient temperature when air veloctiy is 0.1 m/s.

Solution

$$\text{Ambient temperature } t_{am} = t_r = t_a$$

Using Figure 1.5, points on the dashed diagonal line yield equal mean radiant and air temperature. Where the dashed line intersects the isovel (line of constant velocity) of 0.1 m/s, $t_{am} = 23°C$.

It is not usually possible to attain equal mean radiant and air temperatures which are dependent upon the level of insulation of the building, the ventilation rate and the type of heating system. They should, however, be close to this ambient temperature.

Example 1.3
A store room is held at an air temperature of 16°C and 50% relative humidity. The air velocity is 0.1 m/s. A storeman wearing clothing of 1 Clo is allocated to work in the room.

If his activity level is 1 Met find the mean radiant temperature required to provide the right level of comfort.

Solution
From Table 1.3, 1 Met represents a person sitting quietly. It is likely therefore that the storeman is located at a desk or counter.

From the comfort diagram in Figure 1.5, mean radiant temperature $t_r = 33°C$.

This is well outside the second comfort zone used in case study 1.1. However, the whole store does not need radiant heating to this level. A luminous directional radiant heater located at low level would be recommended. It would also be worth suggesting an alternative clothing combination at least to 1.5 Clo. See Table 1.5.

Example 1.4
Determine the conditions that will provide thermal comfort for seated spectators at a sports centre swimming pool.

Their clothing combination is 0.5 Clo and their activity level is 1 Met. Assume that the air velocity is 0.2 m/s and relative humidity is 80%.

Solution
From Table 1.3, 1 Met refers to a person seated quietly and from Table 1.5, 0.5 Clo is equivalent to light summer clothing.

From the comfort diagram in Figure 1.6, air and mean radiant temperature should be 26.2°C.

Example 1.5
Assistants working in a retail store have an estimated activity level of 2 Met. The clothing worn has an insulation level of 1 Clo. If the air velocity is 0.4 m/s find the ambient temperature which should provide thermal comfort for the assistants.

Solution
From Table 1.3, 2 Met indicates medium activity and from Table 1.5, 1 Clo relates to indoor winter clothing.

From the comfort diagram in Figure 1.7, air and mean radiant temperature should be 17.2°C and relative humidity 50%.

Example 1.6
An office is staffed by personnel whose clothing insulation value is 1 Clo and who are engaged in activity estimated at 1 Met. Relative humidity is 50%, air velocity 0.1 m/s and mean radiant and air temperature is 23°C. Installation of new business equipment increases the mean radiant temperature to 26°C.

By what amount should the air temperature controls be changed to maintain thermal comfort?

Solution
From the comfort diagram in Figure 1.5, for a mean radiant temperature of 26°C, required air temperature is 21°C.

This will mean re-setting the controls (23 − 21) = 2 K lower than before.

It is likely that by doing this the mean radiant temperature of 26°C will be lowered and therefore fine tuning may be necessary.

Example 1.7
Factory operatives work in an environment in which the air velocity is 0.5 m/s and relative humidity is 50%. The factory overalls provide an insulation value of 1 Clo and the activity level is expected to be 2 Met.

Recommend suitable ambient conditions for the factory.

Solution
From the comfort diagram in Figure 1.7, air and mean radiant temperature should be 17.5°C.

QUANTIFYING THERMAL SATISFACTION

ASHRAE has developed a seven-point scale of assessment for thermal environments and Professor Fanger has developed a method of predicting the level of satisfaction using the Predicted Mean Vote (PMV). Table 1.6 gives the ASHRAE scale and the corresponding PMV.

Table 1.6 Comfort scales

ASHRAE scale	Fanger PMV index
Hot	3
Warm	2
Slightly warm	1
Neutral	0
Slightly cool	−1
Cool	−2
Cold	−3

Professor Fanger converted the PMV index to the Predicted Percentage of Dissatisfied (PPD) in order to show the PMV as a percentage of people occupying a thermal *environment* who would be likely to be dissatisfied with the level of comfort. Figure 1.8 which is a normal distribution shows the conversion from PMV to PPD. If, for example, the predicted mean vote index is −1 or +1, the predicted percentage dissatisfied, from Figure 1.8, is 26%. The other 74% are likely to be satisfied. When the PMV is zero (Table 1.6) the PPD, from Figure 1.8, is 5%. This implies, for example, that out of 40 people occupying a room in which the PMV is zero, two people are likely to be dissatisfied. The other 95% or 38 people will be satisfied.

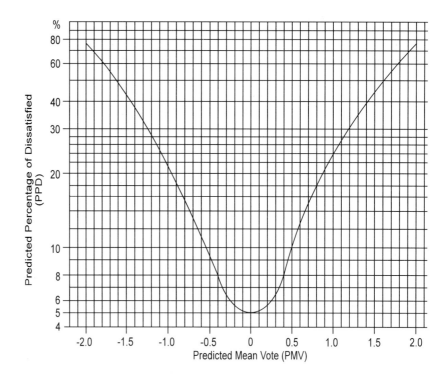

Figure 1.8 Predicted Percentage of Dissatisfied (PPD) as a function of Predicted Mean Vote (PMV).

QUANTIFYING THE LEVEL OF DISSATISFACTION

Clearly it is important to ensure that the level of dissatisfaction with the thermal environment is kept to a minimum, preferably to no more than 5% of the people populating a building.

Table 1.7 allows the PMV index to be obtained for ambient temperatures between 20°C and 27°C and air velocities from < 0.1 to 1.0 m/s when the relative humidity is 50%, the metabolic rate is 1 Met and the clothing insulation level is 1 Clo.

Table 1.8 gives the rate of change of PMV with ambient temperature [d(PMV)/dt_{am}] and Table 1.9 gives the rate of change of PMV with mean radiant temperature [d(PMV)/dt_r].

The four variables to be measured are: dry bulb temperature, relative humidity (wet bulb temperature), mean radiant temperature and air velocity.

The point of measurement is normally taken as the centre in plan and the height above floor is normally 0.6 m for sedentary occupancy and 1.0 m for standing occupancy.

Table 1.7 Predicted Mean Vote Index

Ambient temperature	RH = 50%, M = 1 Met, I_{clo} = 1 Clo Relative air velocity (U_a) (m/s)							
(t_{amb} °C)	< 0.1	0.1	0.15	0.20	0.30	0.40	0.5	1.00
20	−0.85	−0.87	−1.02	−1.13	−1.29	−1.41	−1.51	−1.81
21	−0.57	−0.60	−0.74	−0.84	−0.99	−1.11	−1.19	−1.47
22	−0.30	−0.33	−0.46	−0.55	−0.69	−0.80	−0.88	−1.13
23	−0.02	−0.07	−0.18	−0.27	−0.39	−0.49	−0.56	0.79
24	0.26	0.20	0.10	0.02	−0.09	−0.18	−0.25	0.46
25	0.53	0.48	0.38	0.13	0.21	0.13	0.07	−0.12
26	0.81	0.75	0.66	0.60	0.51	0.44	0.39	0.22
27	1.08	1.02	0.95	0.89	0.81	0.75	0.71	0.56

Table 1.8 Rate of change of PMV with ambient temperature $\left(\dfrac{dPMV}{dt_{am}}\right)$

Clothing insulation	RH = 50%, M = 1 Met Relative air velocity (U_a)(m/s)			
(I_{cl}) Clo	<0.1	0.2	0.5	1.0
0.5	0.350	0.370	0.435	0.490
1.0	0.258	0.260	0.290	0.310
1.5	0.198	0.200	0.217	0.230

Table 1.9 Rate of change of PMV with mean radiant temperature $\left(\dfrac{dPMV}{dt_r}\right)$

Clothing insulation	RH = 50%, M = 1 Met Relative air velocity (U_a)(m/s)			
(I_{cl}) Clo	<0.1	0.2	0.5	1.0
0.5	0.160	0.155	0.140	0.12
1.0	0.117	0.107	0.090	0.077
1.5	0.090	0.080	0.067	0.055

Example 1.8
A factory has the thermal conditions given below. It is proposed to section off part of the factory for employees clothed at 1 Clo and engaged in sedentary work at 1 Met.

Determine the predicted percentage of dissatisfied for the area and determine the value to which the mean radiant temperature must be raised to provide an acceptable level of thermal comfort.

Data: air temperature 21°C, mean radiant temperature 16°C, air velocity 0.2 m/s, relative humidity 50%.

Solution

From reference to the thermal comfort diagram in Figure 1.5 the intersection of the air and mean radiant temperatures are to the left of the 0.2 m/s velocity isovel. The conditions therefore are too cool and do not produce thermal comfort.

From Table 1.8 ambient temperature is taken as air temperature of 21°C and at an air velocity of 0.2 m/s, $PMV_{ta} = -0.84$.

Since mean radiant temperature is too low $[d(PMV)/dt_r]$ from Table 1.9 is obtained where air velocity is 0.2 m/s and clothing insulation is 1 Clo.

Thus $[d(PMV)/dt_r] = 0.107$

The PMV correction factor $PMV_c = [d(PMV)/dt_r](t_a - t_r)$
$= 0.107(21 - 16) = 0.535$ and $PMV = PMV_{ta} - PMV_c = -0.84 - 0.535 = -1.375$

From Figure 1.8   PPD = 44%

From the comfort diagram in Figure 1.5, given air temperature as 21°C and air velocity as 0.2 m/s, mean radiant temperature is read off as 29°C.

You should now undertake the solution based upon an air velocity of 0.1 m/s.

1.9 Temperature profiles

The vertical temperature profile in the conditioned space will vary with the type of heating/cooling employed to offset the heat losses/gains.

Figure 1.9 shows temperature profiles resulting from different types of space heating. The ideal profile is one which is close to vertical. Low air and mean radiant temperatures at floor level or high temperatures at head level will encourage levels of discomfort.

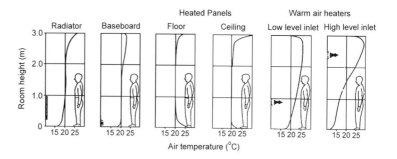

Figure 1.9 Vertical temperature profiles. (Reproduced from the *CIBSE Guide* (1986) by permission of the Chartered Institution of Building Services Engineers.)

1.10 Chapter closure

This chapter has provided you with the underpinning knowledge of the response mechanisms of the human body to surrounding air and mean radiant temperatures and of the thermal comfort criteria which have a direct bearing upon the level of comfort. The thermal indices used to measure these criteria have been identified and used to analyse comfort levels. This should encourage you to pursue, investigate and analyse other situations. Personal thermal comfort is, of course, a subjective matter and therefore it is unlikely that all the occupants of a conditioned building will agree on the level of comfort provided within it. This should not act as a discouragement from attempts to ensure a thermally comfortable environment indoors for the majority of occupants.

2 Heat conduction

Nomenclature

A	surface area (m²)	
A_i	inner surface area	
A_m	logarithmic mean surface area	
A_o	outer surface area	
AWAR	area weighted average thermal resistance	
e	emissivity of surface	
eh_r	heat transfer coefficient for radiation (W/m²K)	
ESP	expanded polystyrene	
h_c	heat transfer coefficient for convection (W/m²K)	
h_s	heat conductance in the surface film (W/m²K)	
h_{si}	surface conductance at the inner surface (W/m²K)	
h_{so}	surface conductance at the outer surface (W/m²K)	
I	heat flux (W/m²)	
k	thermal conductivity (W/mK)	
L	thickness (m)	
Q	heat flow (W)	
Q/L	heat flow per unit length (W/m)	
R	thermal resistance (m²K/W)	
R_a	thermal resistance of the air cavity	
R_c	reciprocal of h_c	
R_e	thermal resistance (m²K/W)	
R_g	thermal resistance (m²K/W)	
R_i	thermal resistance of the added material (m²K/W)	
R_p	thermal resistance of the plaster (m²K/W)	
R_r	reciprocal of eh_r	
R_{si}	thermal resistance at the inside surface	
R_{so}	thermal resistance at the outside surface	
R_t	total thermal resistance	
R_{t1}	transform thermal resistance	
R_{t2}	transform thermal resistance	
R_v	thermal resistance of the void	
t_{ai}	indoor air temperature	
t_{ao}	outdoor air temperature	
t_c	dry resultant temperature	
t_{ei}	indoor environmental temperature	
t_{eo}	outdoor environmental temperature	

t_i indoor temperature
t_o outdoor temperature
t_r mean radiant temperature
t_s surface temperature
U thermal transmittance coefficient (W/m²K)

2.1 Introduction

Heat conduction can occur in solids. It can also occur in liquids and gases in which the vibrating molecules are unable to break free from each other because of the presence of boundary surfaces having a small temperature differential.

The air gap in an external cavity wall of a heated building at normal temperatures will contain still air since the temperature difference between indoors and outdoors is insufficient for the air to generate convection currents resulting from the density difference of the air between the inside surface and the outside surface of the cavity.

If the air gap is increased to more than 75 mm there is sufficient room for the air in the cavity to overcome its own viscosity and the resistance at the inside surfaces of the cavity and the difference in density will encourage the air to convect naturally.

The thermal conductivity k of still air is around 0.024 W/mK making it an excellent thermal insulation. Convected air on the other hand provides very poor thermal insulation. Heat conduction also takes place in films of fluids which occur at boundary surfaces such as in air at the inner and outer surface of the building envelope and on the inside surface of pipes and ducts in which water and air respectively are flowing. In these circumstances the heat conduction is considered either as a surface conductance h_{si}, h_{so} or as a surface thermal resistance R.

The *CIBSE Guide*, section A3 lists typical values of thermal conductivity for different materials employed in the building process. This is reproduced in Table 2.1.

Thermal conductivity is one of the properties of a substance. In liquids and gases it is affected by changes in temperature more than in solids. The thermal conductivity of porous solids is affected by the presence of moisture. This is the reason why in Table 2.1 the outer leaf of a brick wall has a higher thermal conductivity than the inner leaf.

Heat conduction can be considered as taking place radially outwards as in the case of an insulated pipe transporting a hot fluid; in two directions as in the case of a floor in contact with the ground and air; and in one direction as in the case of heat flow at right angles through the external building envelope.

2.2 Heat conduction at right angles to the surface

This mode of heat transfer is commonly associated with that through the building structure (Figure 2.1).

Table 2.1 Properties of materials used in buildings reproduced from the CIBSE Guide (1986) (by permission of the Chartered Institution of Building Services Engineers)

Material	Density (kg/m^3)	Thermal conductivity (W/m K)	Specific heat capacity (J/kg K)
Walls			
(External and Internal)			
Asbestos cement sheet	700	0.36	1050
Absestos cement decking	1500	0.36	1050
Brickwork (outer leaf)	1700	0.84	800
Brickwork (inner leaf)	1700	0.62	800
Cast concrete (dense)	2100	1.40	840
Cast concrete (lightweight)	1200	0.38	1000
Concrete block (heavyweight)	2300	1.63	1000
Concrete block (mediumweight)	1400	0.51	1000
Concrete block (lightweight)	600	0.19	1000
Fibreboard	300	0.06	1000
Plasterboard	950	0.16	840
Tile hanging	1900	0.84	800
Surface Finishes			
External rendering	1300	0.50	1000
Plaster (dense)	1300	0.50	1000
Plaster (lightweight)	600	0.16	1000
Roofs			
Aerated concrete slab	500	0.16	840
Asphalt	1700	0.50	1000
Felt/Bitumen layers	1700	0.50	1000
Screed	1200	0.41	840
Stone chippings	1800	0.96	1000
Tile	1900	0.84	800
Wood wool slab	500	0.10	1000
Floors			
Cast concrete	2000	1.13	1000
Metal tray	7800	50.00	480
Screed	1200	0.41	840
Timber flooring	650	0.14	1200
Wood blocks	650	0.14	1200
Insulation			
Expanded polystyrene (EPS) slab	25	0.035	1400
Glass fibre quilt	12	0.040	840
Glass fibre slab	25	0.035	1000
Mineral fibre slab	30	0.035	1000
Phenolic foam	30	0.040	1400
Urea formaldehyde (UF) foam	10	0.040	1400

Note:
Surface resistances have been assumed as follows:

External walls	$R_{so} = 0.06 \, m^2 \, K/W$
	$R_{si} = 0.12 \, m^2 \, K/W$
	$R_a = 0.18 \, m^2 \, K/W$
Roofs	$R_{so} = 0.04 \, m^2 \, K/W$
	$R_{si} = 0.10 \, m^2 \, K/W$
	$R_a = 0.18 \, m^2 \, K/W$ (pitched)
	$R_a = 0.16 \, m^2 \, K/W$ (flat)
Internal walls	$R_{so} = R_{si} = 0.12 \, m^2 \, K/W$
	$R_a = 0.18 \, m^2 \, K/W$
Internal floors	$R_{so} = R_{si} = 0.12 \, m^2 \, K/W$
	$R_a = 0.20 \, m^2 \, K/W$

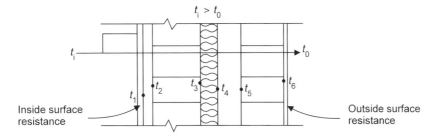

Figure 2.1 Conductive heat flow through a composite wall.

Fourier's law for one-dimensional steady-state heat flow through a single slab of homogeneous material at right angles to the surface is:

$$I = k\,dt/L \text{ W/m}^2$$

where dt = temperature difference across the faces of the slab.
This law has its limitations because we usually work from the temperature indoors to the temperature outdoors and this includes the inside and outside surface resistances R_{si} and R_{so}. Furthermore there is frequently more than one slab of material in the building structure being considered.

A more appropriate generic formula is given below in which indoor temperature t_r, t_{ai}, t_{ei} and t_c are for convenience considered to be equal in value and denoted here as t_i, and outdoor temperature t_{ao} and t_{eo} are considered equal and are denoted here as t_o

$$R_t = (1/h_{si}) + \sum(L/k) + R_a + 1/h_{so} = 1/U \text{ m}^2\text{K/W}$$

Where the thermal resistance R for each element in the composite structure is

$$R = L/k \text{ m}^2\text{K/W}$$

In the context of heat conduction the reciprocal of the surface conductance h_{si}, h_{so} at the inside and outside surfaces is normally used thus:

$$R_{si} = 1/h_{si} \quad \text{and} \quad R_{so} = 1/h_{so}$$

and the generic formula becomes:

$$R_t = R_{si} + \sum(L/k) + R_a + R_{so} = 1/U \text{ m}^2\text{K/W} \tag{2.1}$$

From which the thermal transmittance coefficient or U value for the composite structure which includes surface film resistances is calculated.

It follows that the intensity of heat flow (heat flux) I will be:

$$I = U(t_i - t_o) \text{ W/m}^2 \tag{2.2}$$

Furthermore from equations (2.1) and (2.2) $I = dt/R_t \text{ W/m}^2$

If indoor to outdoor temperature difference dt is steady the heat flux I will be steady. Thus $dt/R_t = I =$ Constant, and therefore

$$dt \propto R$$

so

$$dt_1/R_1 = dt_2/R_2 \, W/m^2 \qquad (2.3)$$

This allows the determination of face and interface temperatures in a composite structure at steady temperatures.

The conductive heat flow through the composite structure may be determined from:

$$Q_s = UA(t_i - t_o) \, W$$

Consider the conductive heat flow path through a composite structure having two structural elements and an air cavity, Figure 2.2.

Equation (2.3) may be adapted as follows:

$$(t_i - t_1)/R_{si} = (t_i - t_o)/R_t \text{ from which } t_1 \text{can be determined.}$$

Similarly: $(t_i - t_2)/(R_{si} + R_1) = (t_i - t_o)/R_t$ from which t_2 can be calculated, and so on. The conductive heat flow through a composite structure may be adapted to determine the heat flow, through an external building envelope, for example, consisting of a number of different structures, e.g. walls, glazing, floor, roof and doors as follows:

$$Q_s = \sum (UA)(t_i - t_o) \, W \qquad (2.4)$$

It is not the purpose of this book to undertake building heat loss calculations which must include also the heat loss due to infiltration and which are conveniently done using appropriate software, the manual method being analysed in *Heating and Water Services Design in Buildings*.

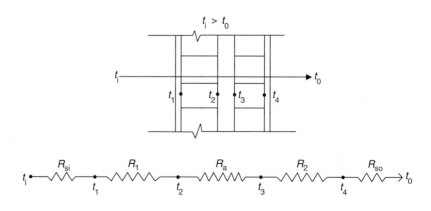

Figure 2.2 Heat flow path through a composite structure.

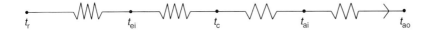

Radiant heating

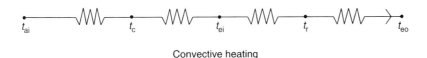

Convective heating

Figure 2.3 Heat flow paths.

However it is appropriate here to point out that in practice the thermal indices t_r, t_c, t_{ei} and t_{ai} are rarely equal in a heated building. This is due to the type of heating system adopted and to an extent upon the level of thermal insulation and natural ventilation. For example, the heat flow path for a building heated by radiant heaters is different to that heated by fan coil units or unit heaters. In the first case the radiant component of heat transfer may be as high as 90% of the total whereas in the second case the convective component of heat transfer will be 100% with no component of heat radiation at all. This clearly has an effect on the heat flow paths through the thermal indices (Figure 2.3). Equation (2.4) therefore will provide an approximate structural heat loss. A more accurate methodology involves the surface conductance h_s.

2.3 Surface conductance

Heat conductance h_s in the surface film combines the coefficients of heat transfer for convection h_c and radiation eh_r thus:

$$h_s = eh_r + h_c \ \text{W/m}^2\text{K}$$

For indoor temperatures around 20°C and using an average value for the heat transfer coefficient for convection and a typical emissivity with the radiation coefficient:

$$h_s = (0.9 \times 5.7) + 3.0 = 8.13 \ \text{W/m}^2\text{K}$$

From which $R_{si} = 1/h_s = 1/8.13 = 0.123 \ \text{m}^2\text{K/W}$

This agrees with the value for R_{si} listed in Table 2.1 and assumes that t_r and t_{ai} are equal. Since this rarely is the case the resulting conductive heat flow must be approximate but is considered good enough for most U value calculations.

A more accurate calculation can be obtained by separating the components of convection and radiation thus:

$$I = eh_r(t_r - t_s) + h_c(t_{ai} - t_s) \ \text{W/m}^2 \qquad (2.5)$$

Example 2.1

A room is held at an air temperature of 20°C when outdoor temperature is −1°C and under steady conditions the inside surface temperature of the external wall is measured at 17.5°C. Assuming a wall emissivity of 0.9 determine:

(i) the rate of conductive heat flow through the wall if $t_r = t_s$;
(ii) the rate of heat flow through the wall when $t_r = t_{ai}$;
(iii) the rate of heat flow through the wall when $t_r = 21°C$.

Solution
(i) $I = eh_r(t_r - t_s) + h_c(t_{ai} - t_s)$
$I = 3(20 - 17.5) = 7.5 \text{ W/m}^2$ of wall.
(ii) $I = 0.9 \times 5.7(20 - 17.5) + 3(20 - 17.5)$
$I = 20.325 \text{ W/m}^2$ of wall.

Also

$$I = h_s((t_r, t_{ai}) - t_s) = 8.13(20 - 17.5) = 20.325 \text{ W/m}^2$$

furthermore

$$I = ((t_r, t_{ai}) - t_s)/R_{si} = (20 - 17.5)/0.123 = 20.325 \text{ W/m}^2.$$

(iii) Clearly in this case the room is radiantly heated for $t_r > t_s$ and
$I = 0.9 \times 5.7(21 - 17.5) + 3(20 - 17.5)$
$I = 25.455 \text{ W/m}^2$ of wall.

Summary for Example 2.1
Note the variations in conductive heat flow:

Case	t_{ai}	t_r	t_s	t_o	Heat flux in W/m^2
(i)	20	17.5	17.5	−1	7.5
(ii)	20	20	17.5	−1	20.325
(iii)	20	21	17.5	−1	25.455

In case (ii) we have the standard method for determining the U value which is when t_r is assumed to be the same as t_{ai}, thus from equation (2.2):

$I = U.dt$ and therefore the external wall $U = I/dt = 20.325/(20 + 1) = 0.968 \text{ W/m}^2\text{K}$. Clearly the heat flux in case (iii) is greater than in case (ii). This will have the effect of increasing the structural heat loss in radiantly heated buildings. The determination of plant energy output Q_p accounts for this whereas the calculation of structural heat loss adopting equation (2.4) may not and therefore may only provide an approximation of structural heat flow at steady temperatures as the summary to Example 2.1 shows. The heat flow paths for the wall are shown in Figure 2.4.

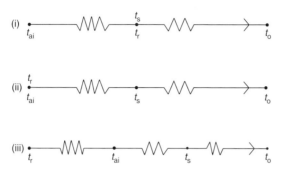

Figure 2.4 Example 2.1 – heat flow paths.

Example 2.2

An external cavity wall is constructed from the components shown below. From the data determine the rate of conductive heat flux at right angles to the surface and calculate the temperature at each face and interface.

Wall construction: inner leaf 10 mm lightweight plaster, 110 mm lightweight concrete block lined on the cavity face with 25 mm of glass fibre slab, air space 50 mm, outer leaf 110 mm brick.

Data: indoor air temperature 23°C, indoor mean radiant temperature 18°C, outdoor temperature −2°C;
thermal conductivities and outside surface resistance are taken from Table 2.1
inside surface convective heat transfer coefficient $h_c = 3.0 \text{ W/m}^2\text{K}$
inside surface heat transfer coefficient for radiation $eh_r = 5.13 \text{ W/m}^2\text{K}$.

Solution

If temperatures remain steady a heat balance may be drawn such that: heat flow from the indoor radiant and air points to the inside surface t_s = heat flow from the inside surface t_s to the outdoor temperature t_o.

Combining equations (2.3) and (2.5) the heat balance becomes:
$eh_r(t_r - t_s) + h_c(t_{ai} - t_s) = (t_s - t_o)/R$ from which t_s may be determined.

Thermal resistance R is taken here from the inside surface to the outside t_o and

$$R = (0.01/0.16) + (0.110/0.19) + (0.025/0.035) + 0.18$$
$$+ (0.110/0.84) + 0.06$$

$$R = 1.7268 \, \text{m}^2\text{K/W}$$

substituting: $5.13(18 - t_s) + 3(23 - t_s) = (t_s + 2)/1.7268$

thus $\quad 159 - 8.86t_s + 119 - 5.18t_s = t_s + 2$

from which $\quad\quad\quad\quad\quad\quad\quad t_s = 18.4°C$

The remaining interface temperatures can now be determined

$$(t_s - t_1)/R_p = (t_s - t_o)/R$$

substituting: $(18.4 - t_1)/0.0625 = (18.4 + 2)/1.7268 = 11.814$
from which $t_1 = 17.66°C$

$$(t_s - t_2)/(R_p + R_c) = (t_s - t_o)/R$$

substituting: $(18.4 - t_2)/(0.0625 + 0.579) = 11.814$ from which $t_2 = 10.82°C$

$$(t_s - t_3)/(R_p + R_c + R_i) = (t_s - t_o)/R$$

substituting: $(18.4 - t_3)/(0.0625 + 0.579 + 0.714) = 11.814$ from which $t_3 = 2.39°C$

$$(t_s - t_4)/(R_p + R_c + R_i + R_a) = 11.814$$

substituting: $(18.4 - t_4)/(0.0625 + 0.579 + 0.714 + 0.18) = 11.814$
from which $t_4 = 0.26°C$

$$(t_5 - t_o)/R_{so} = (t_s - t_o)/R$$

substituting: $(t_5 + 2)/0.06 = 11.814$ from which $t_5 = -1.29°C$

Summary for Example 2.2
The inside surface temperature of the external wall is obtained by separating the heat transfer coefficients for convection and radiation since t_r does not equate with t_{ai}. The heat flow path for the wall is shown in Figure 2.5.

The dew point location in the wall should be checked. It should occur in the outer leaf where vapour can migrate to outdoors. A vapour barrier may be required on the hot side of the thermal insulation, that is at the interface of the plaster and the inner leaf of the wall, to inhibit vapour flow from indoors to outdoors.

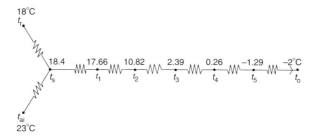

Figure 2.5 Example 2.2 – the heat flow path.

2.4 Heat conduction in ground floors

Heat loss through a solid floor in contact with the ground consists of two components:

edge loss
ground loss

The edge loss is the more significant component and so rooms having ground floors with four exposed edges will have a greater heat loss than rooms with floors having fewer exposed edges.

The formula for the thermal transmittance coefficient U for solid floors in contact with the ground is given as:

$$U = (2k_e B)/(0.5b\pi) \; \text{artanh} \; (0.5b/(0.5b + 0.5w)) \; \text{W/m}^2\text{K} \quad (2.6)$$

where b = breadth (lesser dimension) of the floor in m
w = thickness of surrounding wall taken to be 0.3 m
k_e = thermal conductivity of earth
= 1.4 W/mK

This depends upon the moisture content and ranges from 0.7 to 2.1 W/mK

$$B = \exp(0.5b/L_f) = (2.7183)^{(0.5b/L_f)}$$

L_f = length (greater dimension) of floor in m.

artanh is one of the logarithmic forms of the inverse hyperbolic functions and is expressed in mathematical terms as:

$$\text{artanh}(x/a) = 0.5 \ln((a+x)/(a-x))$$

If another material is added to the composite structure its thermal transmittance U_n can be adjusted thus:

$$U_n = 1/((1/U) + R_i) \; \text{W/m}^2\text{K} \quad (2.7)$$

where R_i is the thermal resistance of the added material.

Example 2.3
A floor in contact with the ground and having four exposed edges measures 20 m by 10 m.

(a) determine the thermal transmittance for the floor;
(b) if the floor is surfaced with 15 mm of wood block having a thermal conductivity of 0.14 W/mK, determine its thermal transmittance coefficient;
(c) if the floor has a 25 mm thermal insulation membrane of EPS having a thermal conductivity of 0.035 W/mK in addition to its wood block finish, determine the transmittance coefficient.

Solution (a)

$$B = \exp(0.5 \times 10/20) = \exp(0.25) = 2.7183^{0.25} = 1.284$$

from the mathematical expression of artanh:

$$\text{artanh}(5/(5+0.15)) = 0.5\ln((5.15+5)/(5.15-5))$$
$$= 0.5\ln(67.667) = 2.1073$$

substituting in equation (2.6)

$$U = ((2 \times 1.4 \times 1.284)/(0.5 \times 10 \times \pi)) \times 2.1073 = 0.482 \text{ W/m}^2\text{K}$$

Solution (b)
From equation (2.7) which accounts for floor finish:

$$U_n = 1/((1/U) + R_i) = 1/((1/0.482) + (0.015/0.14))$$
$$= 1/(2.0747 + 0.107)$$
$$U_n = 0.46 \text{ W/m}^2\text{K}$$

Solution (c)
From equation (2.7)

$$U_n = 1/((1/0.46) + (0.025/0.035)) = 1/(2.174 + 0.714)$$
$$U_n = 0.346 \text{ W/m}^2\text{K}$$

Summary for Example 2.3

floor data	U value
basic	0.482 W/m²K
plus floor finish	0.46
plus floor finish and insulation membrane	0.346

Note:
(i) the floor structure does not play a part in the determination of the U value for floors in contact with the ground. The thermal conductivity of the earth k_e (equation (2.6)) is accounted for.

(ii) the effect of reducing the number of exposed edges from four to two lowers the thermal transmittance by about half here.
(iii) the determination of the heat flux I to ground in practical heat loss calculations is based upon the indoor to outdoor design temperature differential and not indoor to ground temperature.

2.5 Heat conduction in suspended ground floors

A suspended ground floor above an enclosed air space is exposed to air on both sides. The air temperature below the floor will be higher than outdoor air temperature when it is at winter design condition because of the low rate of ventilation under the floor.

The heat flow paths are shown in Figures 2.6a and 2.6b.

Figure 2.6b is the equivalent flow path which can assist in the determination of the thermal transmittance coefficient.

The nomenclature for Figure 2.6 is as follows:

R_g = thermal resistance through floor slab = L/k m²K/W

R_{t1}, R_{t2} = transform resistances from delta to star

$R_{t1} = R_r.R_c/(R_r + 2R_c)$

$R_{t2} = R_c2/(R_r + 2R_c)$

$R_r = 1/eh_r = 1/(0.9 \times 5.7) = 0.2\,\text{m}^2\text{K/W}$

$R_c = 1/h_c = 1/1.5 \qquad = 0.67\,\text{m}^2\text{K/W}$

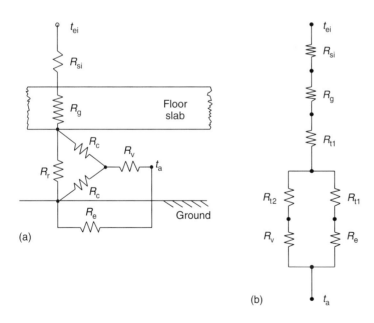

Figure 2.6 (a) Delta arrangement; and (b) star arrangement, of thermal resistance network through suspended ground floors (Example 2.4).

36 Heat conduction

where the heat transfer coefficient h_c downwards at normal temperatures is $1.5 \, W/m^2K$, thus $R_{t1} = (0.2 \times 0.67)/((0.2 + (2 \times 0.67)) = 0.09 \, m^2K/W$

and $R_{t2} = (0.67)2/((0.2 + (2 \times 0.67)) = 0.29 \, m^2K/W$

R_v = ventilation resistance = $0.63b$

$R_e = (1/U) - R_{si}$

where the surface film resistance R_{si} downwards is taken as $0.14 \, m^2K/W$.

The thermal resistance R of the suspended ground floor following the heat flow path in Figure 2.6b can be shown as:

$$R = R_{si} + R_g + R_{t1} + [(1/(R_{t1} + R_e)) + (1/(R_{t2} + R_v))]^{-1} \quad (2.8)$$

Example 2.4
A suspended ground floor consists of 280 mm of cast concrete and measures 20 m by 10 m and has four exposed edges. Determine the thermal transmittance coefficient for the floor.

Solution
$R_g = L/k$ where the thermal conductivity of cast concrete from Table 2.1 is $1.4 \, W/mK$ thus: $R_g = 0.28/1.4 = 0.2 \, m^2K/W$

$R_r = 0.2 \, m^2K/W$ from above

$R_c = 0.67 \, m^2K/W$ from above

$R_{t1} = 0.09 \, m^2K/W$ from above

$R_{t2} = 0.29 \, m^2K/W$ from above

The floor in this example is the same size as that in Example 2.3 and without a floor finish or thermal insulation membrane that U value was calculated as $0.482 \, W/m^2K$.

Make sure you have followed this procedure before continuing. Taking the same conditions for the suspended ground floor in this example to find R_e:

$R_e = (1/U) - R_{si} = (1/0.482) - 0.14 = 1.935 \, m^2K/W$

$R_v = 0.63 \times 10 = 6.3 \, m^2K/W$

$R_{si} = 0.14 \, m^2K/W$ from text above

Substituting these values for the terms in equation (2.8):

$R = 0.14 + 0.2 + 0.09 + [1/(0.09 + 1.935) + 1/(0.29 + 6.3)]^{-1}$

$R = 0.43 + (0.494 + 0.152)^{-1}$

$R = 0.43 + 1.55$

$R = 1.98 \, m^2K/W$

since $U = 1/R = 1/1.98 = 0.505 \, W/m^2K$.

Summary for Example 2.4
Note:
(i) that the thermal transmittance coefficient for the floor in contact with the ground must be determined first. This U value is then used to determine R_e which is in equation (2.8) for suspended ground floors.
(ii) if the floor has a finish of 15 mm thick wood block and a 25 mm thermal insulation membrane of EPS its thermal transmittance can be found adapting equation (2.7)

thus $U_n = 1/((1/U) + R_w + R_i)$

and $U_n = 1/((1/0.505) + (0.015/0.14) + (0.025/0.035))$

from which $U_n = 1/(1.98 + 0.107 + 0.714) = 0.357 \, W/m^2K$.

(iii) the comparison of the transmittance coefficients for the floor in contact with the ground (Example 2.3) and the suspended ground floor, each with four exposed edges can now be made.

Floor structure	suspended ground floor	floor in contact with ground
no insulation or finish	0.505	0.482
with insulation and finish	0.357	0.346

2.6 Thermal bridging and non-standard U values

External walls may not have a thermal transmittance which is consistent over the wall area. Structural columns may form thermal bridges in a cavity wall. At these points the rate of conductive heat flow is high compared with that of the wall. The joisted flat roof is another example where the U value for the joisted part of the roof will be different to that for the spaces between the joists. That part of the structure having the higher U value is usually considered as the thermal bridge and will therefore cause the inside surface temperature to be at a lower value than the rest of the inner surface. Thermal bridges having high U values can cause discoloration of the inside surface and in extreme cases condensation.

There are three types of thermal bridge:

1. Discrete bridges. These include lintels and structural columns which are flush with the wall or take up part of the wall thickness.
2. Multi-webbed bridges which include hollow building blocks.
3. Finned element bridges where the structural column protrudes beyond the width of the wall.

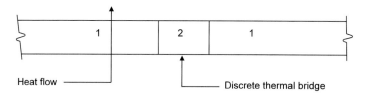

Figure 2.7 A thermally bridged wall.

For walls with discrete bridges the average thermal transmittance

$U = P_1 \cdot U_1 + P_2 \cdot U_2$

where P_1 = unbridged area/total area
and P_2 = bridged area/total area

Refer to Figure 2.7.

For finned element bridges the bridged area in the calculation of P_2 includes the surface area of the protruding part of the thermal bridge.

For a twin leaf wall with a discrete bridge in one of the leaves

$U = 1/(R_b + R_h)$

where R_b = bridged resistance = $1/((P_1/R_1) + (P_2/R_2))$
and $R_1 = R_{si} + L/k + 0.5 R_a$
and $R_2 = R_{si} + L/k + 0.5 R_a$ (bridge)
R_h = homogeneous resistance = $0.5 R_a + L/k + R_{so}$

Refer to Figure 2.8.

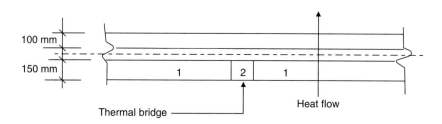

Figure 2.8 Cavity wall with bridge in inner leaf.

Example 2.5
Determine the non-standard U value for the bridged external wall shown in Figure 2.8.

Data: $R_{si} = 0.12$, $R_{so} = 0.06$, $R_a = 0.18 \, m^2 K/W$, $k_b = 0.84$, $k_1 = 0.19$, $k_2 = 1.4 \, W/mK$, $P_1 = 10\%$, $P_2 = 90\%$.

Solution
The data imply an external wall with concrete columns at intervals on the inner leaf.
 Adopting the formulae in the text above:

$R_h = (0.5 \times 0.18) + (0.1/0.84) + 0.06 = 0.269 \, m^2K/W$

$R_1 = 0.12 + (0.15/0.19) + (0.5 \times 0.18) = 0.9995 \, m^2K/W$

$R_2 = 0.12 + (0.15/1.4) + (0.5 \times 0.18) = 0.317 \, m^2K/W$

$R_b = 1/((0.9/0.9995) + (1/0.317)) = 0.822 \, m^2K/W$

substituting for the non-standard U value:

$U = 1/(0.822 + 0.269) = 0.917 \, W/m^2K$

2.7 Non-standard U values, multi-webbed bridges

When considering the thermal transmittance for a hollow block the effect of lateral heat flow is significant. See Figure 2.9 An approximate calculation of the mean thermal resistance involves dividing the hollow block in two planes and employing the area weighted average thermal resistance (AWAR).

Consider the hollow block shown in Figure 2.10. By dividing the block into horizontal sections as shown in Figure 2.10

$R_a = 2L/k$

AWAR $A_1/R_b = (A_2/R_v) + (A_1 - A_2)/(L/k)$

from which R_b is found and

$R_c = R_a + R_b$

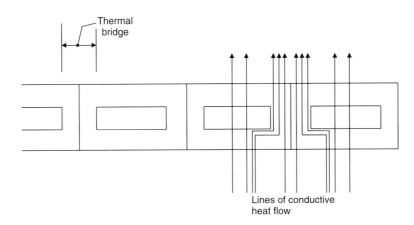

Figure 2.9 Non-uniform heat flow through a hollow block wall.

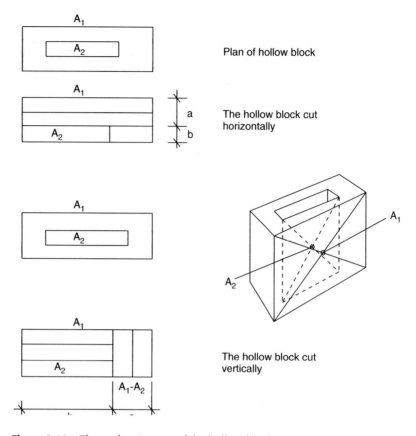

Figure 2.10 Thermal resistance of the hollow block.

By dividing the block into vertical sections as shown in Figure 2.10

$R_a = L/k$

$R_b = (2L/k) + R_v$

AWAR $A_1/R_d = (A_1 - A_2)/R_a + (A_2/R_b)$

from which R_d is found and

$R_m = 0.5(R_c + R_d)$

and the non-standard thermal transmittance $U = 1/(R_{si} + R_m + R_{so})$

Clearly the hollow block will normally form part of the wall structure. If it is rendered on the outside and plastered on the inside, for example, the non-standard thermal transmittance will be

$U = 1/(R_{si} + R_p + R_m + R_r + R_{so})$

Where R_p and R_r are the thermal resistances of the plaster and rendering respectively.

Example 2.6

Figure 2.11 shows an external hollow block wall rendered on the outer face and plastered on the inner face. From the data determine the non-standard thermal transmittance coefficient for the wall and hence the heat flux, given indoor temperature is 20°C and outdoor temperature is −5°C.

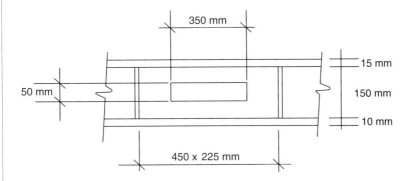

Figure 2.11 Hollow block wall (Example 2.6).

Data: External wall specification: 15 mm of external rendering, medium weight hollow concrete block with air cavity filled with EPS, 16 mm of lightweight plaster. Thermal and film properties taken from Table 2.1.

Solution

You should now refer to Figure 2.10 which shows the way in which the block is cut for the purposes of determining the two thermal resistances R_c and R_d.

It is also important to identify the calculation procedure for the hollow block in the text above.

Slicing the block horizontally we have:

$R_a = 2L/k = 2 \times 0.05/0.51 = 0.196$

Hollow block dimensions and face and void surface areas:

$A_1 = 450 \times 225 = 101\,250$ this is approximately equivalent to 100
$A_2 = 350 \times 225 = 78\,750$ this is approximately equivalent to 79

thus AWAR $A_1/R_b = (A_2/R_v) + (A_1 - A_2)/(L/k)$
Note that R_v is not a void/air cavity here as it is filled with EPS

$100/R_b = (79/(0.05/0.035)) + ((100 - 79)/(0.05/0.51))$

from which $R_b = 0.371$

now $R_c = R_a + R_b = 0.196 + 0.371 = 0.567 \, \text{m}^2\text{K/W}$

Slicing the block vertically we have:

$R_a = L/k = 0.15/0.51 = 0.294$

$R_b = (2L/k) + R_v = (0.1/0.51) + (0.05/0.035) = 1.6246$

Note again that R_v in this case is not a void/air cavity as it is filled with EPS

$$\text{AWAR} \quad A_1/R_d = ((A_1 - A_2)/R_a) + A_2/R_b$$
$$100/R_d = ((100 - 79)/0.294) + (79/1.6246)$$

from which $R_d = 0.833 \, \text{m}^2\text{K/W}$

now $R_m = 0.5(R_c + R_d) = 0.5(0.567 + 0.833) = 0.7 \, \text{m}^2\text{K/W}$
for the composite wall $R = R_{si} + R_p + R_m + R_r + R_{so}$

$R = (0.12) + (0.01/0.16) + (0.7) + (0.015/0.5) + (0.06)$

$R = 0.9725 \, \text{m}^2\text{K/W}$

The non-standard U value for the wall will therefore be:

$U = 1/R = 1/0.9725 = 1.03 \, \text{W/m}^2\text{K}$

The conductive heat flux

$I = Udt = 1.03 \times (20 + 5) = 25.75 \, \text{W/m}^2$ of wall

Summary for Example 2.6
This is an average rate of heat flow through the wall. Thermal bridges are formed between the cavities filled with EPS and may result in discoloration on the inside surface of the plaster due to surface temperature variations along the wall.

A thermal bridge is a path located through a structure where the rate of heat flow is substantially increased as a result of the materials used.

2.8 Radial conductive heat flow

For plane (flat) surfaces, surface area A is constant and $Q = UAdt$ W. For cylinders and spheres surface area is not constant either for multiple layers of material or for single layers having a measurable thickness. Thus as the radius increases through the thermal insulation material surrounding a pipe transporting hot or chilled water, for example, so does the surface area of the insulation surrounding that pipe.

If A_m = the mean surface area of each layer of thermal insulation around the pipe, then from Fourier's equation for a single layer:

$Q = kA_m dt/L$ W

This can be rewritten as

$Q = dt/(L/kA_m)$

Radial conductive heat flow

Thus for multiple layers of insulation around a pipe:

$$Q = dt / \sum (L/kA_m) \, W$$

Where dt = temperature differential between the inside pipe surface and the outside insulation surface.

If dt is taken from the fluid flowing inside the pipe to the outside air:

$$Q = dt / ((1/A_i h_{si}) + \sum (L/kA_m) + (1/A_o h_{so})) \quad (2.9)$$

A_m is the logarithmic mean area of each layer of material and A_i and A_o are the inside and outside surface areas respectively.

For cylinders

$$A_m = (\pi(d_2 - d_1)L)/\ln(d_2/d_1) \quad (2.10)$$

You can see here the similarity with the surface area of a cylinder $A = \pi L d$.

Furthermore A_m is approximately equal to $\pi((d_1 + d_2)/2)L$. You should confirm this in Example 2.7 below.

For spheres

$$A_m = \pi d_1 d_2 \text{ for each thermal insulation layer} \quad (2.11)$$

This is one of the two methodologies for introducing radial heat flow in pipes and circular ducts having one or more layers of thermal insulation.

Example 2.7

A cylindrical vessel 4 m diameter and 7 m long has hemispherical ends giving it an overall length of 11 m. The vessel which stores water for space heating at 85°C is covered with 300 mm of lagging which has a thermal conductivity of 0.05 W/mK. Determine the heat loss from the vessel to the plant room which is held at 22°C.

Take the outside surface heat transfer coefficient as 12 W/m²K and ignore the influence of the vessel wall and the inside heat transfer coefficient.

Solution

Since the inside heat transfer coefficient and the vessel wall thickness is not accounted for, equation (2.9) for cylinders needs adapting and:

$$Q = dt/(L/(kA_m) + (1/h_{so} A_o))$$

Where here $dt = (85 - 22)$ and from equation (2.10)

$$A_m = \pi(4.6 - 4.0)7/\ln(d_2/d_1) = 94.38 \, m^2$$

$$A_o = \pi \times 4.6 \times 7 = 101.16 \, m^2$$

Substituting we have:

$$Q = (85 - 22)/(0.3/0.05 \times 94.38) + (1/(12 \times 101.16))$$

$$Q = 63/(0.0636 + 0.0008)$$

You can see that the effect of the last term involving h_{so} and A_o is insignificant, thus for the cylinder $Q = 978$ W.

The hemispherical ends of the vessel form a sphere, for the logarithmic mean area A_m of which, equation (2.11) can be used and:

$$A_m = \pi \times 4.0 \times 4.6 = 57.8 \text{ m}^2$$

Ignoring the effect of h_{so}, the heat loss from the hemispherical ends of the vessel will be:

$$Q = dt/(L/kA_m)$$

substituting:

$$Q = (85 - 22)/(0.3/0.05 \times 57.8) = 607 \text{ W}$$

The total heat loss from the vessel to the plant room $= 978 + 607 = 1585$ W.

The classical methodology for developing a formula for radial heat flow integrates Fourier's law for one-dimensional heat flow.

Thus: $Q = -kAdt/dr$ W for a single layer of material.

Considering unit length of a cylinder L where $A = 2\pi r L$

$$Q/L = -k(2\pi r)dt/dr \text{ W/m run}$$

If temperatures are steady Q remains steady and the temperature gradient from the fluid flowing in the pipe to surrounding air decreases with increasing radius r.

Integrating between the limits r_1 and r_2, refer to Figure 2.12:

$$\int_1^2 (Q/L)dr/r = -\int_1^2 \pi k dt$$

$$(Q/L)\ln(r_2/r_1) = -2\pi k(t_1 - t_2)$$

From which $Q/L = -2\pi k(t_1 - t_2)/\ln(r_2/r_1)$ W/m for a single layer of material. The minus sign indicates a heat loss and may be ignored.

For a multilayer cylindrical wall:

$$Q/L = (2\pi dt)/((1/r_1 h_{si}) + ((\ln r_2/r_1)/k_2) \\ + ((\ln r_3/r_2)/k_3) \ldots + (1/(r_n h_{so})) \text{ W/m run} \quad (2.12)$$

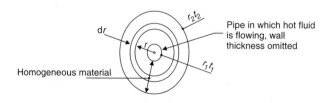

Figure 2.12 Radial heat flow integrating between the limits of r_1 and r_2.

Example 2.8

A 50 m length of steam pipe connects two buildings and carries saturated steam at 29 bar gauge. The pipe has an internal diameter of 180 mm, a wall thickness of 19 mm and is covered by two layers of thermal insulation. The inner layer is 20 mm thick and the outer layer is 25 mm thick. The thermal effect of the outer protective casing to the pipe insulation can be ignored.

Determine the rate of heat loss from the pipe.

Data: outdoor temperature 5°C; thermal conductivity of pipe wall $k_w = 48$, inner layer of insulation $k_i = 0.035$, outer layer of insulation $k_o = 0.06$ W/mK; heat transfer coefficient at the pipe inner surface $h_{si} = 550$, heat transfer coefficient at the outer layer insulation surface $h_{so} = 18$ W/m²K.

Solution
Refer to Figure 2.13. $r_1 = 90, r_2 = 109, r_3 = 129$ and $r_4 = 154$ mm from the tables of *Thermodynamic and Transport Properties of Fluids* the temperature of saturated steam at 30 bar abs. = 234°C. Substituting data into equation (2.12)

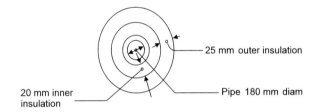

Figure 2.13 Insulated steam pipe (Example 2.8).

$$Q/L = 2\pi(235 - 5)/[(1/0.9 \times 550) + ((\ln 109/90)/48) \\ + ((\ln 129/109)/0.035) + ((\ln 154/129)/0.06) \\ + (1/0.154 \times 18)]$$

Note the ratios of r_2/r_1 etc. are kept in millimetres for convenience without loss of integrity.

$$Q/L = 1439/(0.002 + 0.004 + 4.813 + 2.952 + 0.361)$$
$$Q/L = 1439/8.132 = 177 \text{ W/m run}$$
$$Q = 177 \times 50 = 8847 \text{ W}$$

Summary for Example 2.8
You can see that the effect of the inside heat transfer coefficient h_{si} and the pipe wall is insignificant and therefore frequently ignored in the solution to this type of problem.

Example 2.9

Determine from the data the thickness of thermal insulation to be applied to a pipe conveying water over a distance of 100 m if its temperature is not to fall below 79°C

Data:
outdoor temperature	−1°C
initial water temperature	80°C
water flow rate	1.5 kg/s
outside diameter of the pipe	66 mm
specific heat capacity of water	4200 J/kg/K
thermal conductivity of the insulation	0.07 W/mK
coefficient of heat transfer at the outside surface of the insulation	10 W/m²K

Solution

The effect of the heat transfer coefficient at the inside surface of the pipe can be ignored. The effect of the thickness of the pipe wall may be ignored. The temperature of the outer surface of the pipe can therefore be taken as 80°C at the beginning of the pipe run and 79°C at the end of the run.

Now the maximum heat loss from the pipe will be:

$$Q/L = MCdt/L = (1.5 \times 4200 \times (80 - 79))/100 = 63 \text{ W/m run}$$

This is equivalent to a heat loss of $63/(\pi dL)$ W/m²

$$= 63/(\pi \times 0.066 \times 1) = 304 \text{ W/m}^2$$

The mean temperature of the pipe surface $= (80 + 79)/2 = 79.5$ °C

Let $z =$ insulation thickness in metres

Adopting equation (2.12)

$$63 = (2\pi(79.5 + 1))/[\ln((0.033 + z)/(0.033)/0.07) + (1/10(0.033 + z)].$$

Collecting the common factors on the left hand side and inverting the formula:

$$161\pi/63 = (\ln(0.033 + z)/0.07) + 1/10(0.033 + z).$$

This may be written as $8.03 = Y + W$

If values are now given to z then the terms Y and W in the formula can be reduced to numbers. The results are given in Table 2.2.

From the tabulated results in Table 2.2 it can be observed that the insulation thickness lies between 16 and 18 mm for the equation $8.03 = Y + W$ to balance. If z and $(Y + W)$ are plotted on a graph the thickness may be obtained as 17.3 mm. In practice 20 mm thick pipe insulation would no doubt be selected.

Table 2.2 Results for solution to Example 2.9

z	0.033 + z	Y	Z	Y + W
0.01	0.043	3.75	2.33	6.08
0.012	0.045	4.44	2.22	6.66
0.014	0.047	5.07	2.13	7.20
0.016	0.049	5.65	2.04	7.69
0.018	0.051	6.22	1.96	8.18
0.020	0.053	6.78	1.89	8.67

Summary for Example 2.9
1. The maximum heat loss from the pipe is 63 W/m run. This may be converted to a permitted heat loss expressed in W/m^2:

$$I = 63/(\pi dL) = 63/(\pi \times 0.066 \times 1.0) = 304 \, W/m^2$$

2. It is suggested that you now undertake a similar calculation to find the minimum thickness of thermal insulation for the pipe given a maximum permitted heat loss of $200 \, W/m^2$ when outdoor temperature is $-1°C$.

The following start may help:

$$200 \, W/m^2 = (200 \times \pi \times 0.066 \times 1.0) = 41.47 \, W/m \text{ run of pipe}$$

since $Q/L = MCdt/L \quad dt = ((Q/L) \times L)/MC$

thus $dt = (41.47 \times 100)/(1.5 \times 4200) = 0.66 \, K$.

This means that the temperature of the pipe/water at the end of the run will be $80 - 0.66 = 79.34°C$ and the mean temperature will be $(80 + 79.34)/2 = 79.67°C$. This data can now be substituted into the formula and values given to z, the insulation thickness. The solution comes to 36 mm of pipe insulation.

Example 2.10
A sheet steel duct 600 mm in diameter carries air at $-25°C$ through a room held at $20°C$ db and 68% saturated to a food freezing processor. Determine the minimum thickness of thermal insulation required to prevent condensation occurrence on the outer surface of the duct insulation. Recommend a suitable finish to the duct insulation.

Data: atmospheric pressure 101 325 Pa; heat transfer coefficient at the outer surface $8 \, W/m^2K$; thermal conductivity of duct insulation $k = 0.055 \, W/mK$. Ignore the effect of the heat transfer coefficient at the inner surface of the duct and the duct thickness.

48 Heat conduction

Solution
From hygrometric data dew-point temperature t_d of air at 20°C db and 68% saturated is 14°C. To avoid the incidence of condensation the outer surface of the thermal insulation must not fall below this temperature.

A heat balance may be drawn up, assuming steady temperatures, and:

the heat flow from the room to the outer surface of the duct insulation = the heat flow from the insulation outer surface to the air in the duct.

Let z = minimum thickness of insulation in metres and adopting the classical formula (2.12) for radial heat flow and generic equation (2.3), modified for the heat balance.
Thus:

$$2\pi(t_i - t_{so})/(1/h_{so}(0.3 + z)) = 2\pi(t_{so} - t_a)/(\ln((0.3 + z)/0.3)/k)$$

simplify and substitute:

$$(20 - 14)/(1/8(0.3 + z)) = (14 + 25)/(\ln((0.3 + z)/0.3)/0.055)$$

rearranging the equation: $6 = ((39/(\ln((0.3 + z)/0.3)/0.055))$
$$(1/8(0.3 + z))$$

this is a simple relationship of $6 = Y \times W$

Giving values to insulation thickness z and tabulating in Table 2.3.

Table 2.3 Results for solution to Example 2.10

z	(0.3 + z)	Z	Y	YW
0.02	0.32	0.391	33.23	12.99
0.03	0.33	0.379	22.51	8.53
0.04	0.34	0.368	17.40	6.31
0.045	0.345	0.362	15.35	5.55

Summary for Example 2.10
From Table 2.3 it is seen that the minimum thickness of duct insulation is about 42 mm. The surface finish to the duct insulation must be waterproof to ensure against the migration of water or vapour into the thermal insulation from the surrounding air.

2.9 Chapter closure

You have now been introduced to the principles of conductive heat flow and applied these principles to practical applications relating to:

heat flow through ground floors
heat flow resulting from non-standard U values
radial heat flow from thermally insulated pipes, ducts and vessels.

You are familiar with the potential errors in the determination of building heat loss when inequalities between indoor air temperature and mean radiant temperature are ignored, and the fact that errors are most likely where buildings are not thermally insulated or sealed to current Building Regulation standards.

3 Heat convection

Nomenclature

C	specific heat capacity (kJ/kgK)
d, d	characteristic dimension (m)
D	characteristic dimension (m)
dt	temperature difference (K)
Gr	Grashof number
h_c	heat transfer coefficient for convection (W/m^{2}K)
h_{si}	heat transfer coefficient at the inside surface (W/m^{2}K)
h_{so}	heat transfer coefficient at the outside surface (W/m^{2}K)
I	heat/cooling flux (W/m^{2}K)
k	thermal conductivity kW/mK (W/mK)
M	mass flow rate (kg/s)
Nu	Nusselt number
Pr	Prandtl number
Re	Reynolds number
R_f	fouling resistance (m^{2}K/W)
T	absolute temperature (K)
T_1	absolute temperature (K)
T_2	absolute temperature (K)
T_a	absolute temperature (K)
t_f	customary temperature (°C)
T_m	absolute temperature (K)
T_s	absolute temperature (K)
t_s	customary temperature (°C)
u	mean velocity (m/s)
U	thermal transmission coefficient (W/m^{2}K)
v	specific volume (m^3/kg)
x, x	characteristic dimension (m)
β	coefficient of cubical expansion (T^{-1})
μ	dynamic viscosity (kg/ms)
γ	kinematic viscosity (m^2/s)
ρ	(kg/m^3)

3.1 Introduction

Heat convection can occur in liquids and gases when the molecules move freely and independently. This occurs as a result of cubical expansion or contraction of the fluid when it is heated or cooled,

Introduction 51

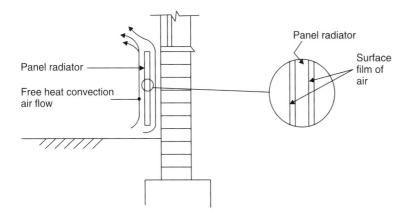

Figure 3.1 Air flow in free convection over panel radiator.

causing changes in fluid density, which initiates movement by natural means with the warmer fluid rising and the cooler fluid dropping, due to the effects of gravity upon it. This type of movement is termed free convection and occurs over radiators and natural draught convectors for example (Figure 3.1).

Forced convection is obtained with the aid of a prime mover such as a pump or fan and occurs at the heat exchanger of a fan coil unit, for example, with pumped water flowing inside the heat exchanger pipes and fan assisted air flowing over the finned heat exchanger surface. See Figures 3.2 and 3.3.

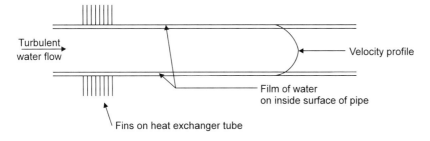

Figure 3.2 Water flow inside finned tube heat exchanger providing forced convection.

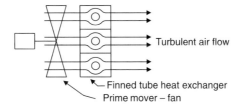

Figure 3.3 Air flow in forced convection over a finned tube heat exchanger.

52 Heat convection

Free convection relies on natural forces and its effectiveness in heat transfer relies on:

> velocity of the secondary fluid over the heat exchanger surface
> magnitude of the temperature difference between the primary and secondary fluids
> size and shape of heat exchanger and its position in space. Refer to equations (3.5) to (3.9).

The velocity of the secondary fluid over the heat exchanger surface will influence:

- type of flow whether laminar or turbulent
- the leaving temperature of the secondary fluid off the heat exchanger
- the effect of the film at the interface of the moving secondary fluid and the heat exchanger surface on the convective heat transfer
- the degree of contact between the secondary fluid and the heat exchanger surface.

Forced convection on the other hand is less affected by the shape of the heat exchanger or its position in space. It has a positive and directional movement and is not so subject to natural forces. However most applications of forced convection rely on it as the main mode if not the only mode of heat transfer. In the case of unit heaters or fan coil units in heating mode, which are mounted at high level/ceiling level, it is important to account for the fact that leaving air temperatures make the air buoyant and this buoyancy must be overcome to ensure that it reaches the working plane.

There are a number of properties of the fluid flowing which influence heat convection:

$\beta, \mu, \gamma, \rho, k, C, T$.

There are other quantities which are affected or which affect heat convection:

M, dt, h_c, u.

These variables together with appropriate characteristic dimensions may be collected in dimensionless groups by analysis as shown here. Refer also to Chapter 11.

Reynolds number	$Re = \rho u d / \mu = dM/\mu A = ud/\gamma$	(3.1)
Nusselt number	$Nu = h_c d / k$	(3.2)
Prandtl number	$Pr = \mu C / k$	(3.3)
Grashof number	$Gr = \beta(\rho^2)(x^3)dtg/\mu^2$	(3.4)

Notes:
(i) cubical expansion of gases $\beta = 1/T_m$ K^{-1}
where $T_m = (T_s + T_a)/2$

(ii) properties of air and water at different temperatures can be obtained from the tables of *Thermodynamic and Transport Properties of Fluids* (SI units) and include: C, k, μ, ρ, v, Pr density $\rho = 1/v$
(iii) before adopting a formula for the purposes of solving a problem it is necessary to determine whether the fluid is in laminar flow or turbulent flow. Refer to Chapter 6 for a detailed analysis.

3.2 Rational formulae for free and forced heat convection

The formulae given in this section have been determined by dimensional analysis (Chapter 11) and the associated constants and indices have been determined empirically which is to say by practical experiment. The solutions resulting from application of these formulae must therefore be treated as approximate but sufficiently accurate.

Because of the complexity of determining the heat transfer coefficient for convection h_c from a surface, specific formulae have been developed for different shapes of surface in various positions in space. Some of the more appropriate of these are given here.

The following Grashof and Reynolds numbers identify the type of fluid flow and associated formulae are given.

Free convection of air over vertical plates:

| Laminar flow | $Gr < 10^8$ |
| Turbulent flow | $Gr > 10^9$ |

$$\text{Laminar flow} \quad (Nu)_x = 0.36(Gr)_x^{0.25} \tag{3.5}$$

$$\text{Turbulent flow} \quad (Nu)_x = 0.13((Pr)(Gr)_x)^{0.33} \tag{3.6}$$

Free convection of air over horizontal plates:

Laminar flow $\quad 1.4 \times 10^5 < (Gr) < 3 \times 10^7$

Turbulent flow $\quad 3 \times 10^7 < (Gr) < 3 \times 10^{10}$

For hot surfaces looking up and cool surfaces looking down:

$$\text{Laminar flow} \quad h_c = 1.4((t_s - t_f)/D)^{0.25} \tag{3.7}$$

For hot surfaces looking down and cool surfaces looking up:

$$\text{Laminar flow} \quad h_c = 0.64((t_s - t_f)/D)^{0.25} \tag{3.8}$$

where $\quad D = (\text{length} + \text{width})/2$

For hot surfaces looking up and cool surfaces looking down:

$$\text{Turbulent flow} \quad h_c = 1.7(t_s - t_f)^{0.33} \tag{3.9}$$

Forced convection.
Turbulent flow inside tubes where $Re > 2500$

$$(Nu)_d = 0.023(Re)d^{0.8}(Pr)^{0.33} \tag{3.10}$$

Turbulent flow outside tube bundles where $Re > 4 \times 10^5$

$$(Nu)_d = 0.44(Re)d^{0.55}(Pr)^{0.31} \tag{3.11}$$

Turbulent flow over flat plates where $Re > 1 \times 10^5$

$$(Nu)_x = 0.037(Re)x^{0.8}(Pr)^{0.33} \tag{3.12}$$

Free convection over horizontal cylinders.
Laminar flow where $Gr < 10^8$ for air and water

$$(Nu)_d = 0.53((Gr)_d(Pr))^{0.25} \tag{3.13}$$

Free convection over vertical cylinders.
Turbulent flow where $Gr > 10^9$

for air $\quad Nu = 0.1((Gr)(Pr))^{0.33}$ (3.14)

for water $\quad Nu = 0.17((Gr)(Pr))^{0.33}$ (3.15)

From the above text you will have seen that:

free convection is related to the Grashof number Gr where laminar and turbulent flow depends upon its magnitude.
forced convection is related to the Reynolds number Re where laminar and turbulent flow depends upon its magnitude.

From the Nusselt formula the heat transfer coefficient for convection h_c is determined and convective heat transfer is obtained from:

$$Q = h_c \times A \times dt \text{ W} \tag{3.16}$$

where dt = surface temperature t_s minus fluid temperature t_f
and A = area of heat exchanger surface

The heat transfer coefficient for convection h_c is dependent upon the magnitude of the difference between the heat exchanger surface temperature and the bulk fluid temperature and also upon the thickness of the laminar sublayer on the heat exchanger surface. Refer to Chapter 6.

3.3 Temperature definitions

Various terms are used for temperature and temperature difference in convective heat transfer. They include:

- Mean bulk temperature which refers to the arithmetic mean temperature of the fluid flowing.
- Mean film temperature which refers to the mean temperature of the bulk plus heat exchanger temperature.
- Mean temperature difference refers to the difference between heat exchanger and mean bulk temperature.
- Log mean temperature difference (LMTD). If the temperature of both fluids vary, true temperature difference will be the logarithmic mean value.

Example 3.1
Given a rise in temperature of the fluid flowing of 70°C to 80°C as a result of a constant heat exchange temperature of 140°C, find the mean bulk temperature, the mean film temperature of the fluid and the mean temperature difference.

Solution
A heat exchanger whose surface is at a constant temperature is usually associated with the use of steam or refrigerant which gives up its latent heat in the heat exchanger at constant temperature and in the process changes its state.

Mean bulk temperature of the fluid flowing $= (70 + 80)/2 = 75°C$
Mean film temperature of the fluid flowing $= 0.5((70 + 80)/2 + 140) = 107.5°C$

Mean temperature difference between the fluid flowing and the heat exchanger surface $= (140 - (70 + 80)/2) = 65\,K$

Example 3.2
Given a constant fluid temperature of 20°C and a heat exchanger surface temperature of 110°C determine the mean bulk temperature, mean film temperature and mean temperature difference.

Solution
This is unusual since the heat given up by the fluid flowing must result in a change in state at constant temperature as it does also with the fluid in the heat exchanger.

Mean bulk temperature of the fluid flowing = 20°C
Mean film temperature of the fluid flowing $= 0.5(20 + 110) = 65°C$
Mean temperature difference between the fluid and the heat exchanger $= (110 - 20) = 90\,K$

Example 3.3
A counterflow heat exchanger carries high temperature hot water at inlet and outlet temperatures of 150°C and 110°C. The secondary fluid rises in temperature from 70°C to 82°C. Find the logarithmic mean temperature difference between the primary and secondary fluids.

Solution
This topic is considered in detail in Chapter 9. Log mean temperature difference dt_m accounts for temperature variations in both the primary and secondary fluids and from Chapter 9:

$$dt_m = (dt_{max} - dt_{min})/\ln(dt_{max}/dt_{min})$$

for counterflow: primary 150 $\longrightarrow$ 110
secondary 82 $\longleftarrow$ 70
$\overline{\quad 68 dt_{max} \quad 40 dt_{min} \quad}$

substitute: $dt_m = (68 - 40)/\ln(68/40) = 52.77\,\text{K}$

Summary for Example 3.3
It is interesting to note that the arithmetic mean $= [((150 + 110)/2) - ((82 + 70)/2)] = (130 - 76) = 54\,\text{K}$ which is not much different to the logarithmic mean.

If the high temperature return is at 120°C instead of 110°C the logarithmic mean temperature difference is calculated as 58.5 K whereas the arithmetic mean is 59 K. Here the two values are even closer. You should now confirm that this is so.

If conductive heat flow is considered from the hot fluid across the exchanger wall to the cold fluid the overall U value for the heat exchanger is appropriate:

$$U = (1/((1/h_{si}) + R_f + (1/h_{so}))) \quad \text{W/m}^2\,\text{K} \tag{3.17}$$

hence

$$Q = UA\,dt \quad \text{W} \tag{3.18}$$

where $dt =$ temperature of hot fluid minus the temperature of the cold fluid.

Note:
(i) The thermal resistance of the heat exchanger wall is ignored since it is insignificant.
(ii) The overall U value is dependent upon the thickness of the laminar sublayer on both the inside and the outside of the heat exchanger and hence on the type of fluid flow. See Chapter 6.
(iii) R_f is the fouling factor measured in m²K/W. If maintenance of the heat exchanger is done on a regular basis it is sometimes ignored. It ranges from 0.000 09 to 0.0002 m²K/W.

There now follows some practical examples which adopt the formulae introduced in the text above. You should follow them through noting the solution procedure in each case.

3.4 Convective heat output from a panel radiator

Example 3.4
Low temperature hot water flows through a single panel vertically mounted radiator 1 m high by 1.5 m long. The mean temperature of the circulating water is 60°C and the temperature of the surroundings air is 19°C.

Determine the heat transferred by convection. Ignore the resistance of the air and water films on each side of the radiator and the radiator material. Evaluate the properties at the mean film temperature.

Solution
The first step in the solution procedure is to determine the Grashof number which will establish whether the air flow over the radiator is in the laminar or turbulent region.

The mean film temperature $= 0.5(60 + 19) = 39.5°C = 312.5\,K$

Referring to the Grashof number, equation (3.4)

cubical expansion $\beta = 1/(312.5) = 0.0032$

Interpolating from the tables of *Thermodynamic and Transport Properties of Fluids*, air density at mean absolute temperature of 313 K

$\rho = 1.13\,kg/m^3$

and air viscosity $\mu = 0.000\,019\,kg/ms$
vertical height $x = 1.0\,m$
temperature difference between the radiator and air $dt = 60 - 19 = 41\,K$.
Now substituting into the Grashof number:

$Gr = 0.0032 \times (1.13)^2 \times (1.0)^3 \times 41 \times 9.81/(0.000\,019)^2$

$Gr = 4.55 \times 10^9$

Note that x is the height of the panel; a lower height yields a lower Gr and ultimately a lower heat transfer coefficient h_c. A long low radiator will therefore give a lower convective output than a short tall radiator of the same area. This is caused by the increased stack effect of the taller radiator, which induces greater vertical air flow over its surface.

From the text, turbulent flow in free convection over vertical plates commences when $Gr > 10^9$ and therefore applies here. The adopted formula will be equation (3.6).

The Prandtl number can be calculated or the value interpolated from the tables of *Thermodynamic and Transport Properties of Fluids* at the mean air temperature of 313 K in which case

$$Pr = 0.703$$

Note that the tables quote the thermal conductivity k in kW/mK and specific heat capacity C in kJ/kgK. By interpolating μ, C and k for dry air from the tables Pr can also be calculated from equation (3.3).

from equation (3.6) $(Nu)_x = 0.13((Pr)(Gr)_x)^{0.33}$
substituting: $(Nu)_x = 0.13((0.703)(4.55 \times 10^9))^{0.33}$
$(Nu)_x = 178$

But from equation (3.2) $(Nu)_x = h_c x/k$ where x is a characteristic dimension and here x is the panel height thus $x = 1.0$
and

$$h_c = Nuk/x = 178 \times 0.0273/1.0 = 4.86 \text{ W/m}^2\text{K}$$

the surface area of the radiator is $1.0 \times 1.5 \times 2 = 3.0 \text{ m}^2$
from equation (3.16) free heat convection $Q = h_c A dt = 4.86 \times 3 \times (60 - 19) = 598 \text{ W}$

Summary for Example 3.4
As already mentioned in the solution the height x of a radiator influences its heat transfer by free convection by affecting the Grashof number. The lower the radiator height the lower is the convective heat transfer for the same surface area. This is confirmed from reference to manufacturers' literature. In practical terms the 'stack effect' of the free convection over the radiator surface increases with its height thus increasing convective output.

3.5 Heat output from a pipe coil freely suspended

Example 3.5
A 100 mm bore horizontal pipe freely suspended is located at low level in a greenhouse to provide heating. It has a surface emissivity of 0.9 and is supplied with water at 85°C flow and 73°C return. The greenhouse is held at an air temperature of 19°C and a mean radiant temperature of 14°C. Evaluating the properties at the mean film temperature determine the heat output from the pipe given that it is 15 m in length. Take the outside diameter as 112 mm.

Solution
The Grashof number will determine the type of air flow over the pipe.

The mean film temperature $= 0.5[(85+73)/2 + 19] = 49\,°C = 322\,K$

Using the data for dry air from the tables for *Thermodynamic and Transport Properties of Fluids*, and the question:
$C = 1.0063\,\text{kJ/kgK}$, $\mu = 0.00001962\,\text{kg/ms}$, $\rho = 1.086\,\text{kg/m}^3$, $k = 0.00002816\,\text{kW/mK}$, $dt = 79 - 19 = 60\,K$, $x = d = 0.112\,m$
also $\beta = 1/T_m = 1/322 = 0.0031$

Substituting into equation (3.4)

$Gr = (0.0031 \times (1.086)^2 \times (0.112)^3 \times 60 \times 9.81)/(0.00001962)^2$

$Gr = 7.87 \times 10^6$

This identifies laminar air flow over the pipe and from the tables at the mean film temperature of 322 K

$Pr = 0.701$

alternatively it can be determined from equation (3.2)

$Pr = \mu C/k = 0.00001962 \times 1.0063/0.00002816$

from which

$Pr = 0.702$

As air flow is laminar and convection is free equation (3.13) can be adopted thus:

$Nu = 0.53(7.87 \times 10^6 \times 0.701)^{0.25} = 25.7$

from equation (3.3) $h_c = Nuk/d = 25.7 \times 0.02816/0.112 = 6.46\,\text{W/m}^2\text{K}$

From equation (3.16) $Q = 6.46 \times (\pi \times 0.112 \times 15)(79 - 19) = 2046\,W$

The calculation of heat radiation (equation (4.12) and $I = Q/A$) can be made from:

$Q = \sigma e_1 (T_1^4 - T_2^4) A_1$

substituting:

$Q = 5.67 \times 10^{-8} \times 0.9((352)^4 - (287)^4)(\pi \times 0.112 \times 15)$

$Q = 2307\,W$

The total output of the pipe coil $= 2046 + 2307 = 4353\,W$

3.6 Heat transfer from a tube in a condensing secondary fluid

Example 3.6
(a) Calculate the convective heat transfer coefficient at the inside surface of a 24 mm diameter tube in which pumped water is flowing at 0.5 kg/s given water flow and return temperatures of 15°C and 25°C. Evaluate the properties at the mean bulk temperature.
(b) If ammonia vapour at a pressure of 15.54 bar is condensing on the outside surface of the tube which is 30 mm outside diameter, determine the surface area of the tube. Take the heat transfer coefficient at the outer surface of the tube for the condensing ammonia as $10\,kW/m^2K$ and ignore the temperature drop through the tube wall.

Solution (a)
To establish that the flow of water is turbulent the Reynolds number can be used and from equation (3.1) $Re = dM/\mu A$

The mean bulk temperature of the water flowing $= (15 + 25)/2 = 20\,°C$

From the tables for *Thermodynamic and Transport Properties of Fluids* the following properties of water are obtained:

$C = 4.183\,kJ/kgK$, $\mu = 0.001\,002\,kg/ms$, $k = 0.000\,603\,kW/mK$, $Pr = 6.95$

adopting equation (3.1)

$$Re = 0.024 \times 0.5 \times 4/0.001\,002\pi(0.024)^2$$
$$Re = 26\,473$$

since $Re > 2500$ the water flow inside the tube is in the turbulent region and as the water is pumped equation (3.10) for forced convection can be adopted thus:

$$(Nu)_d = 0.023(Re)d^{0.8}(Pr)^{0.33}$$

substitute: $\quad Nu = 0.023(26\,473)^{0.8}(6.95)^{0.33}$
from which $\quad Nu = 151$
from equation (3.2) $\quad Nu = h_c d/k$
substituting: $\quad h_c = 151 \times 0.000\,603/0.024 = 3.794\,kW/m^2\,K$

Solution (b)
From equation (3.17) the overall U value for the heat exchange tube can be determined; and $U = 1/((1/h_{si}) + R_f + (1/h_{so}))$ assuming negligible fouling resistance $R_f =$ zero. As the heat transfer coefficient for radiation eh_r is negligible at the inside surface of the tube $h_{si} = h_c = 3.794\,\text{kW/m}^2\,\text{K}$ and h_{so} is given here as $10\,\text{kW/m}^2\text{K}$.

Substituting into equation (3.17) $U = 1/((1/3.794) + (1/10)) = 2.75\,\text{kW/m}^2\,\text{K}$

From the tables for *Thermodynamic and Transport Properties of Fluids* ammonia at 15.54 bar absolute has a saturation temperature of 40°C.

Adopting the heat balance to determine the net surface area of the heat exchange tubes:

heat gain to the water = the heat loss from the ammonia.

Since the heat loss from the ammonia $Q = UAdt$ from equation (3.18)

then $MCdt$ (water) $= UAdt$ (ammonia)

substituting:

$$0.5 \times 4.183 \times (25 - 15) = 2.75 \times A \times (40 - 20)$$

from which

$$A = 0.38\,\text{m}^2$$

Now tube length $L = A/\pi d = 0.38/\pi \times 0.03 = 4.03\,\text{m}$ of heat exchange tubing.

Summary for Example 3.6
You will have noticed the temperature differential between the ammonia and the bulk temperature of the water. Here it is 20 K. The rate of heat transfer is largely influenced by this temperature differential. If the temperature differential between the primary and secondary fluids is below 15 K heat transfer is poor.

The efficiency of heat exchange is also dependent upon the extent of the contact of the fluid with the outer surface of the heat exchanger. The contact factor identifies the extent of this contact which is, among other things, dependent upon the velocity of the fluid over the surface of the heat exchanger. See Chapter 9.

3.7 Cooling flux from a chilled ceiling

Example 3.7
A chilled beam ceiling operates using chilled water at a mean bulk temperature of 8°C in a room held at an air temperature of 20°C by means of a system of displacement ventilation, and mean radiant temperature is 17°C. Determine the cooling flux in W/m² of chilled ceiling surface which has an emissivity of 0.9.

Solution
The first step is to determine the Grashof number which will establish the type of air flow over the ceiling surface.

The mean film temperature at the ceiling surface will be $0.5(20 + 8) = 14\,°\text{C} = 287\,\text{K}$

Interpolating the properties of dry air at 287 K from the tables for *Thermodynamic and Transport Properties of Fluids*:

$\rho = 1.233\,\text{kg/m}^3$, $\mu = 0.00001783\,\text{kg/ms}$
$\beta = 1/T_m = 1/287 = 0.0035$

Let $x = 1$, $dt = (20 - 8)$
substitute into equation (3.4)

$$Gr = 0.0035 \times (1.233)^2 \times 1^3 \times (20 - 8) \times 9.81/(0.00001783)^2$$

from which $Gr = 1.98 \times 10^9$

For turbulent flow $1.4 \times 10^5 < (Gr) < 3 \times 10^{10}$ and therefore air flow over the ceiling surface is turbulent and free, equation (3.9) applies and $h_c = 1.7(t_s - t_f)^{0.33}$
substituting:

$$h_c = 1.7(20 - 8)^{0.33} = 3.86\,\text{W/m}^2\text{K}$$

the convective heat transfer $Q = h_c A dt$

thus the cooling flux by free convection $Q/A = I = h_c dt = 3.86 \times (20 - 8)$

from which

$I = 46.32\,\text{W/m}^2$ of chilled ceiling surface.

The cooling flux by heat radiation is obtained from equation (4.12) thus:

$$I = \sigma e_1 (T_1^4 - T_2^4)\,\text{W/m}^2$$
$T_1 = 273 + 17 = 290\,\text{K}$ and $T_2 = 273 + 8 = 281\,\text{K}$

and substituting

$$I = 5.67 \times 10^{-8} \times 0.9(290^4 - 281^4)$$

from which

$I = 42.77$ W/m^2 of chilled ceiling surface.

The total cooling flux to the chilled ceiling $I = 46.32 + 42.77 = 89.1$ W/m^2.

Summary for Example 3.7
Dimension x was given a value of 1.0 in the solution for the Grashof number as the cooling flux is for an area of 1 m × 1 m of chilled ceiling surface. As the air flow over the ceiling is in the turbulent region the final solution will not be affected if the cooling flux is now applied to a given area of chilled ceiling.

Had the air flow been in the laminar region equation (3.7) would apply in which case $D =$ (length + width)$/2$ and this clearly affects the value of the heat transfer coefficient for convection h_c. The value of x in the determination of the Grashof number can again be taken as 1.0 without serious error.

3.8 Heat flux off a floor surface from an embedded pipe coil

Example 3.8
(a) Determine the heat flux from the surface of a floor in which a hot-water pipe coil is embedded in a room held at an air and mean radiant temperature of 20°C. The average floor temperature is 26°C and its emissivity is 0.9.
(b) Calculate the conductance h_s at the floor surface under the conditions in part (a).

Solution
(a) The Grashof number will be evaluated first to establish the type of air flow over the floor. The mean bulk temperature is $(26 + 20)/2 = 23\,°\text{C} = 296$ K.

From the tables for *Thermodynamic and Transport Properties of Fluids*, the following properties of dry air at 296 K are obtained:
$C = 1.0049$ kJ/kgK, $\mu = 0.000\,018\,46$ kg/ms, $\rho = 1.177$ kg/m^3,
$\beta = 1/T_m = 1/296 = 0.003\,38$
Substitute into the Grashof formula equation (3.4)

$$Gr = 0.003\,38 \times (1.177)^2 \times 1^3 \times (26 - 20)$$
$$\times 9.81/(0.000\,018\,46)^2$$

from which $Gr = 8.14 \times 10^8$
for turbulent flow $3 \times 10^7 < (Gr) < 3 \times 10^{10}$

Thus flow of air over the floor surface is in the turbulent region and free, and equation (3.9) applies:

$$h_c = 1.7(26 - 20)^{0.33} = 3.071 \text{ W/m}^2\text{K}$$

the free convective heat flux

$$I = h_c dt = 3.071 \times (26 - 20) = 18.42 \text{ W/m}^2$$

the heat radiation flux from the floor surface, using the appropriate equation (4.12)

$$I = \sigma e_1(T_1^4 - T_2^4) = 5.67 \times 10^{-8} \times 0.9((299)^4 - (293)^4)$$
$$= 31.77 \text{ W/m}^2$$

the combined heat flux from the floor $I = 18.42 + 31.77 = 50.2 \text{ W/m}^2$.

(b) By adapting equation (4.13) the heat transfer coefficient for radiation eh_r at the floor surface will be:

$$eh_r = I/dt = 31.77/(26 - 20) = 5.295 \text{ W/m}^2\text{K}$$

the heat transfer coefficient for free convection h_c at the floor surface is calculated as:

$$h_c = 3.071 \text{ W/m}^2\text{K}$$

Since air and mean radiant temperature are equal, the coefficients can be combined and surface conductance h_s will be:

$$h_s = 5.295 + 3.071 = 8.366 \text{ W/m}^2\text{K}.$$

Summary for Example 3.8
The solution to part (b) falls within the range of 8 to 10 W/m²K of published values for the combined heat transfer coefficient at the floor surface for embedded pipe coils. The maximum floor surface temperature should not exceed 27°C to avoid discomfort.

Example 3.9
A single pass shell and tube condenser is required to condense refrigerant at 45°C. It contains 10 tubes each having an internal diameter of 24 mm and an external diameter of 30 mm. Cooling water is available at 10°C and total flow is 6.0 kg/s.

Given that the temperature of the cooling water leaving the condenser is not to exceed 20°C and ignoring the temperature drop through the tube wall, determine the minimum tube length required.

Take the heat transfer coefficient of the condensing vapour as 6.4 kW/m²K. The water side heat transfer coefficient should be determined from the appropriate rational equation.

Solution
Fluid flow in heat exchangers is discussed in some detail in Chapter 9. It can be assumed here that the water flow is subject to a prime mover and thus convection is forced. The Reynolds number Re can be calculated to establish turbulent flow conditions inside the tubes. From the tables for *Thermodynamic and Transport Properties of Fluids*, condenser water at a bulk temperature of $(20 + 10)/2 = 15\,°C = 288\,K$ has the following properties:
$v = 0.001\,m^3/kg$, $\quad C = 4.186\,kJ/kgK$, $\quad \mu = 0.001\,136\,kg/ms$,
$k = 0.595\,W/mK$, $Pr = 7.99$.

From these properties $\rho = 1/0.001 = 1000\,kg/m^3$.
Now volume flow rate $= u \times \pi d^2/4 = Mv \quad m^3/s$.

Rearranging the formula for volume flow rate in terms of mean velocity, then mean velocity in each tube

$$u = (4Mv/\pi d^2) = 4 \times (6.0/10) \times 0.001/\pi \times (0.024)^2 = 1.326\,m/s$$

substitute into equation (3.1)

$$Re = 1000 \times 1.326 \times 0.024/0.001\,136 = 28\,014.$$

The minimum value for Re for turbulent flow inside tubes is 2500 and therefore flow is in the turbulent region here and equation (3.10) applies:

$$Nu = 0.023(28014)^{0.8}(7.99)^{0.33}$$

from which $Nu = 165$
from equation (3.2) $h_c = 165 \times 0.595/0.024 = 4091\,W/m^2K$.

The overall heat transfer coefficient U across the tubes is obtained from equation (3.17) and ignoring the effect of the fouling resistance R_f,

$$U = 1/((1/6.4) + (1/4.09)) = 2.495\,kW/m^2K$$

The output from each tube in the condenser is calculated from $Q = MCdt$

thus $Q = (6.0/10) \times 4.2 \times (20 - 10) = 25.2\,kW$

The overall heat transfer for each tube in the condenser is obtained from equation (3.18)

thus $Q = U(\pi dL)dt = 2.495 \times (\pi \times 0.03 \times L) \times (45 - 15)$.

Using a heat balance in which:

heat gain by condenser water = heat loss by refrigerant

then by substitution:
$$25.2 = 2.495 \times \pi \times 0.03 \times L \times 30$$
from which tube length $L = 3.57\,\text{m}$

Summary for Example 3.9
No account has been taken here of the efficiency of heat exchange which is largely dependent upon the contact factor of the condensing refrigerant on the heat exchange tubes and upon the primary to secondary temperature difference which here is 30 K and well above the minimum of 15 K.

3.9 Chapter closure

You now have practical skills and the underpinning knowledge relating to the application of heat transfer by free and forced convection. The examples and solutions will have given some insight into the application of the rational formulae in use for this mode of heat transfer at steady temperatures, and the procedure for attempting problem solving. Further work on heat exchangers is done in Chapter 9.

Cross-referencing with Chapter 11 is required for the origins of the dimensionless groups employed in this mode of heat transfer, and Chapter 6 for a detailed analysis of laminar and turbulent flow.

Heat radiation 4

Nomenclature

a	absorbtivity
A	surface area of radiator/receiver (m^2)
c	velocity of wave propagation (m/s)
C_1	constant (Wµm^4/m^2)
C_2	constant (µmK)
C_3	constant (µmK)
e	emissivity
$eh_r, (F_{1,2})h$	heat transfer coefficient for radiation from a grey body (W/m^{2}K)
exp	(e^x)
f	function of
$F_{1,2}$	form factor, view factor
h_r	heat transfer coefficient for radiation from a black body (W/m^{2}K)
I	heat flux, intensity of monochromatic radiation, intensity of heat radiation exchange, solar constant (W/m^2)
p_{rt}	plane radiant temperature (°C)
Q	rate of energy flow (W)
r	reflectivity
S	distance between radiator and receiver (m)
t	transmissivity
T	absolute temperature (K)
t_a	air temperature (°C)
t_c	dry resultant temperature/comfort temperature (°C)
t_r	mean radiant temperature (°C)
U	thermal transmittance coefficient (W/m^{2}K)
v	frequency (Hz cycles/s)
v_{rt}	vector radiant temperature (°C)
λ	wavelength (µm)
σ	Stefan–Boltzmann constant of proportionality (W/m^{2}K^4)

4.1 Introduction

A simple definition of heat radiation would be the interchange of electromagnetic waves between surfaces of differing temperatures which can see each other. In fact the full definition is extensive and complex and requires substantial initial briefing and qualification before recourse can be made to practical applications.

All surfaces which are above absolute zero ($-273.15°C$ or $0.0\,K$) are emitting radiant heat or absorbing, reflecting and transmitting heat radiation depending upon whether they are emitting surfaces or receiving surfaces and upon whether the material is opaque or transparent. The distinction between an emitting surface and a receiving surface is dependent upon its temperature in relation to other surfaces it can 'see'.

One of the major differences between heat radiation exchange and that of heat conduction and heat convection is that it does not require an exchange medium.

The sun transfers its heat by radiation through space to the Earth's atmosphere through which it passes to the surface of the Earth. The ozone layer, atmospheric particles, condensing water vapour and dust act as filters to solar radiation reducing its intensity at the Earth's surface. The effect of dust and other particles in the air has the same filtering effect for radiant space heaters which results in a reduction in heat flux at the receiving surfaces.

If the heat radiator cannot 'see' the surfaces to be heated the effects of heat radiation are not immediately apparent, if at all.

Heat radiation is part of the spectrum of light ranging from the ultraviolet to the infrared region – that is from short wave to long wave radiation. Space heating equipment which provides heat radiation includes luminous heaters having temperatures up to $2200°C$ which emit short-wave radiation, and non-luminous heaters which emit invisible long-wave radiation. Examples of each include the electric quartz heaters and radiant strip and, of course, the ubiquitous panel radiator both of which emit heat radiation which can be felt but not seen.

4.2 Surface characteristics

Heat radiation travels in the same wave patterns as light – see Figure 4.1 – and at the speed of light which is 2.98×10^8 m/s, frequently taken as $c = 3 \times 10^8$ m/s. The effectiveness of radiation exchange is dependent upon the texture of the radiating and receiving surfaces. Surface characteristics include:

reflectivity r
transmissivity t
absorptivity a
emissivity e

If a mirror receives heat radiation it will reflect about 97%, thus $r = 0.97$ and its ability to absorb radiation will be about 3%, thus $a = 0.03$. Its surface temperature therefore will not rise by much.

Most receivers will reflect and absorb proportions of the incident radiation. A perfect radiator will emit 100% of its radiation. The sun is a perfect radiator (known as 'black body' radiation) but most surfaces are not and a radiator having a matt surface for example will emit about 90% of its radiation thus $e = 0.9$.

Surface characteristics 69

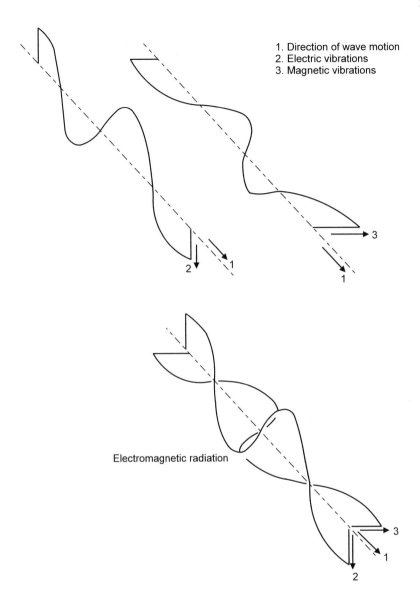

Figure 4.1 Transverse wave motion of electric and magnetic fields vibrating in phase at the same frequency at right angles in a plane perpendicular to the direction of travel.

Kirchhoff established that for most surfaces the ability to emit and absorb heat radiation at the same absolute temperature is approximately equal, thus $e = a$ (equation (4.4)).

For most surfaces $r = (1 - a) = (1 - e)$ and therefore $a = (1 - r)$ and $e = (1 - r)$.

The transmission t of radiant heat through a material occurs by heat conduction resulting from the temperature rise induced by the radiant

70 Heat radiation

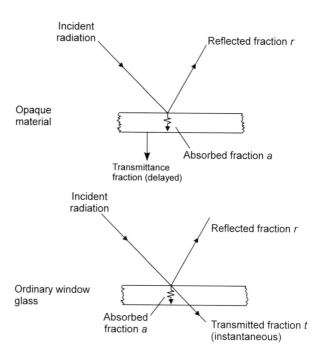

Figure 4.2 The effects of incident heat radiation on a surface.

heat incident upon the surface. It is therefore similar to the thermal transmittance coefficient U. Refer to Figure 4.2.

If a material which is transparent like ordinary window glass, is irradiated the short-wave radiation will be transmitted with little absorption or reflection. Table 4.1 gives some typical values for emissivity, absorptivity and reflectivity for some opaque surfaces. The table gives a general indication only since for many materials emissivity varies with temperature.

Table 4.1 Surface characteristics relating to incident heat radiation

Material/surface	Emissivity e	Absorptivity a	Reflectivity r
Brick and stone	0.9	0.9	0.1
Aluminium, polished	0.04	0.04	0.96
Aluminium, anodized	0.72	0.72	0.28
Cast iron	0.8	0.8	0.2
Copper, polished	0.03	0.03	0.97
Copper, oxidized	0.86	0.86	0.14
Galvanized steel	0.25	0.25	0.75
Paint, metal based	0.5	0.5	0.5
Paint, gloss white	0.95	0.95	0.05
Paint, matt black	0.96	0.96	0.04

Colour temperature indicators

It is useful to have an approximate feel for the temperature of luminous radiant heaters. Table 4.2 gives a rough guide.

Table 4.2 Colour temperature guide for luminous heaters

Colour	Temperature (°C)
Very dull red	500–600
Dark blood red	600–700
Cherry red	700–800
Bright red	800–900
Orange	900–1000
Yellow	1000–1100
Yellow/white	1100–1200
White	1200–1300

4.3 The greenhouse effect

Greenhouse glass allows the transmission of short-wave solar radiation but disallows long-wave transmission. The resultant effect is known as the greenhouse effect. As the surfaces within the greenhouse warm up due to the incidence of short-wave solar radiation passing through the glass, they begin to emit long-wave radiation which cannot escape resulting in a rise in temperature within the greenhouse. This is used to good effect for the propagation of plants and as a means of passive space heating in the winter.

4.4 Spectral wave forms

The pattern of solar radiation follows a sine wave (Figure 4.3). The spectral proportions in which heat radiation is present are given in Table 4.3.

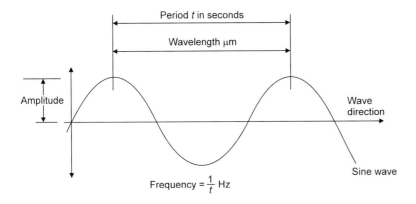

Figure 4.3 Heat radiation waveform.

72 Heat radiation

Table 4.3 Spectral proportions of heat radiation

Cosmic, gamma and X-ray wavebands, wavelength 10^{-11} to 10^{-2} μm
9% ultraviolet, invisible short-wave, wavelength 0.29 to 0.4 μm
40% visible light, short-wave, wavelength 0.40 to 0.7 μm
51% invisible infrared, long-wave, wavelength 0.70 to 3.5 μm
Radio, TV and radar wavebands, wavelength 100 μm to 10^5 m

A discussion on heat radiation needs to identify three types of emitting and receiving opaque surface in order to reduce the complexity of the subject. These are black, grey and selective. A 'black' surface is that of a perfect radiator in which emissivity is constant at any wavelength. Black body (perfect) radiation is rare. It is approached within a boiler furnace and with luminous radiant heaters but not often elsewhere in the real world. Selective surfaces are those of every day manufactured or natural materials. The emissivity of selective surfaces varies arbitrarily with wavelength. This makes it difficult to integrate the radiant heat flux over all the wavelengths for a given temperature of a selective emitter or receiver. A grey surface is an imaginary surface in which the emissivity varies uniformly with wavelength making it easier to integrate the radiant heat flux over all the wavelengths for the given surface. Since the grey surface has a constant heat flux ratio with that of a black body, $I\lambda/Ib\lambda$, it irons out the arbitrary nature of the selective surface making it easier to determine the heat flux I.

4.5 Monochromatic heat radiation

Heat radiation emitted at any one wavelength is called monochromatic radiation. Figure 4.4 shows the variation in black body emissive

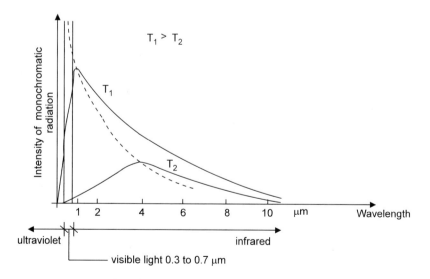

Figure 4.4 Variation of black body emissive power with wavelength and absolute temperature.

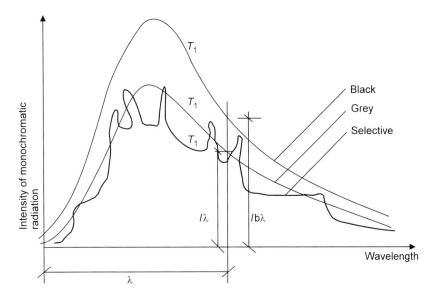

Figure 4.5 Comparison of emission from black, grey and selective surfaces at a fixed temperature T_1.

power with wavelength and absolute temperature for two bodies at temperature T_1 and T_2. Note how wavelength increases as maximum absolute temperature of the body decreases. This was identified by Wein and is known as Wein's displacement law. See equation (4.5). It is denoted by a uniform line declining towards the right on the graph.

The area under each of the curves at T_1 and T_2 represents the sum of the monochromatic radiations or total heat radiation from the surfaces.

A comparison of emissions from black, grey and selective surfaces is shown in Figure 4.5. Note that the ratio of $I\lambda/Ib\lambda$ is constant for the grey surface and arbitrary for the selective surface where it varies at each wavelength. Note also that the absolute temperature of each surface is the same. It is the monochromatic emissive power from each surface which varies with wavelength.

4.6 Laws of black body radiation

The following laws apply to heat radiation from a perfect radiator: Kirchhoff's law and Stefan's law

$$\text{heat flux } I \propto T^4 \text{ W/m}^2 \tag{4.1}$$

The Stefan–Boltzmann constant of proportionality $\sigma = 5.67 \times 10^{-8}$ W/m²K⁴

Thus heat radiation for the sum of the wavelengths

$$I = \sigma T^4 \text{ W/m}^2 \tag{4.2}$$

74 Heat radiation

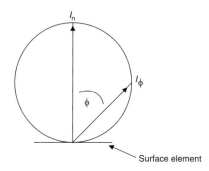

Figure 4.6 A surface element does not radiate energy with equal intensity in all directions, $I_\phi = I_n \cos\phi$.

Planck's law shows the relationship between the monochromatic emissive power I at wavelength λ and absolute temperature T, thus

$$I = C_1 \lambda^{-5} / \exp(C_2/\lambda T) - 1 \text{ W/m}^2 \tag{4.3}$$

Kirchhoff's law $e = a$ \hfill (4.4)

Wein's displacement law $\lambda_{max} T = C_3 \, \mu\text{m K}$ \hfill (4.5)

Wein's wavelength law $\lambda = c/v$ m \hfill (4.6)

$\left.\begin{array}{l}\text{Lambert's law for emissive}\\\text{power from a flat radiator}\end{array}\right\} I\phi = \ln\cos\phi$ \hfill (4.7)

See Figure 4.6.

Stefan–Boltzmann law for heat radiation exchange over the sum of the wavelengths at temperature T_1

$$I = \sigma(T_1^4 - T_2^4) \text{ W/m}^2 \tag{4.8}$$

The constants above have the following numerical values:

c = velocity of wave propagation = 3×10^8 m/s

$C_1 = 3.743 \times 10^8$ Wμm^4/m^2

$C_2 = 1.4387 \times 10^4$ μmK

$C_3 = 2897.6$ μmK

4.7 Laws of grey body radiation

When radiating surfaces remain grey in a system of heat exchange, the emissivities of those surfaces must be accounted for as well as their geometric configuration. In general the heat exchange by radiation between two surfaces will depend upon:

- relative areas of surfaces
- geometry of the surfaces in relation to each other
- the two emissivities.

These factors which are quite complex are identified as the form/view factor $F_{1,2}$ and $F_{1,2} = f(A_1, A_2, e_1, e_2)$

The view factor was introduced in 1951 by H.C.Hottell as a means of accounting for the emissivities of the surfaces and their geometric configuration. The *CIBSE Guide* tabulates form factors for various surfaces' configurations in section C3.

In view of the complexity of determining the form factor, only three applications will be considered here.

Two parallel grey surfaces in which $A_1 = A_2$ and $T_1 > T_2$.

A typical application here is the radiation exchange between the two inside surfaces of a cavity wall.

$$F_{1,2} = 1/((1/e_1) + (1/e_2) - 1). \tag{4.9}$$

Concentric cylindrical surfaces in which $A_1 < A_2$ and $T_1 > T_2$.

A typical application is layers of thermal insulation on cylindrical ducts and pipes.

$$1/F_{1,2} = (1/e_1) + (A_1/A_2)((1/e_2) - 1). \tag{4.10}$$

A small radiator contained in a large enclosure in which $A_1 < A_2$ and $T_1 > T_2$

$$F_{1,2} = e_1.$$

This relationship is approximately correct for most applications of space heating and cooling in which A_2 is the surface area of the enclosing space and T_2 its area weighted mean surface temperature or mean radiant temperature.

Thus for grey body heat radiation

$$I = (F_{1,2})\sigma(T_1^4 - T_2^4) \, \text{W/m}^2 \tag{4.11}$$

4.8 Radiation exchange between a grey body and a grey enclosure

Clearly an imaginary surface whose characteristics iron out the variations in emissivity at different wavelengths and which follows a constant heat flux ratio is more easily analysed than the random variations in emissivity with wavelength which occur with a selective surface. Consider heat radiation exchange between a small grey body radiator located in a grey enclosure as shown in Figure 4.7.

Assuming the absorptivity of the enclosure is 0.9 and that the radiator emits 100 units of heat radiation. Of the 100 units, 99.9 are absorbed by the enclosure so the apparent absorptivity of the surrounding enclosure boundaries is 0.999 Although this is a simplified analysis of the matter it can be argued that the boundary surfaces, initially taken as grey, approach that of a black body enclosure.

Thus from equation (4.11) and the notes in section 4.7

$$I = \sigma e_1(T_1^4 - T_2^4) \, \text{W/m}^2 \tag{4.12}$$

where T_1 is the absolute temperature of the radiator and T_2 is the absolute area weighted mean radiant temperature of the surrounding

76   Heat radiation

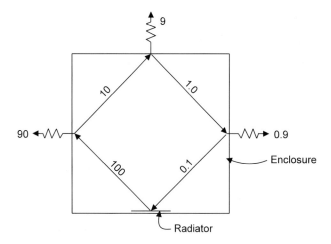

Figure 4.7 Heat radiation exchange. If the enclosure has an absorptivity of 0.9 then out of 100 units of radiant heat 99.9 are absorbed.

surfaces which is approximately equal to the mean radiant temperature of the enclosure T_r.

4.9 Heat transfer coefficients for black and grey body radiation

The heat transfer coefficient for convection h_c, chapter 3, equation (3.16), is:

$$Q = h_c \times A \times dt \; W$$

thus for heat convection $I = h_c(t_1 - t_2) = h_c(T_1 - T_2) \text{W/m}^2$.

For black body heat radiation exchange $I = h_r(T_1 - T_2) \text{ W/m}^2$.

Thus

$$h_r = I/dT \; \text{W/m}^2\text{K} \tag{4.13}$$

The heat transfer coefficient for radiation can also be determined from above where $I = h_r(T_1 - T_2) \text{ W/m}^2$

and from equation (4.8) $I = \sigma(T_1^4 - T_2^4) \text{ W/m}^2$

equating these two formulae $h_r = \sigma(T_1^4 - T_2^4)(T_1 - T_2)^{-1}$

expanding $h_r = \sigma(T_1^2 - T_2^2)(T_1^2 + T_2^2)(T_1 - T_2)^{-1}$

and $h_r = \sigma(T_1 - T_2)(T_1 + T_2)(T_1^2 + T_2^2)(T_1 - T_2)^{-1}$

from which for black bodies

$$h_r = \sigma(T_1 + T_2)(T_1^2 + T_2^2) \; \text{W/m}^2\text{K} \tag{4.14}$$

For grey bodies

$$(F_{1,2})h_r = \sigma(F_{1,2})(T_1 + T_2)(T_1^2 + T_2^2) \tag{4.15}$$

for a small radiator in a large enclosure

$$(e_1)h_r = \sigma(e_1)(T_1 + T_2)(T_1^2 + T_2^2) \tag{4.16}$$

Note that in equations (4.15) and (4.16) ($F_{1,2}$) and (e_1) do not cancel in the determination of the heat transfer coefficient for radiation for grey bodies.

4.10 Heat radiation flux I

As the distance S between the receiving surface and the emitting surface increases, the radiation flux is less intense. This is borne out by varying the distance a person is with respect to a radiant heater like a luminous electric fire. Three applications are considered here.

(i) Point source radiation: the direction of intensity is spherical here hence A_2 is the surface area of a sphere and $A_2 = 4\pi S^2$. If the radius is doubled – that is to say if the distance S from the point source radiator is doubled – the enclosing receiving area A_2 is quadrupled.

Thus, if originally the distance between the point source radiator and the enclosing area is 3, $S = 3$ and $A_2 = 4\pi 3^2 = 36\pi$ whereas if S is doubled to 6, $A_2 = 4\pi 6^2 = 144\pi$ which is four times larger. The effect upon radiation flux or intensity is a four-fold reduction. The total heat radiation received by the enclosing surface of radius 3 is the same as that for the enclosing surface of radius 6, however.

Thus $I \propto 1/S^2$.

(ii) Line source radiator: here $I \propto 1/S$

An example might be a single ceramic luminous rod heater with no reflector and considering radiant heat flux in one plane only.

(iii) A surface element: does not radiate energy with equal intensity in all directions and Lambert's law applies, equation (4.7). See Figure 4.6.

It identifies the fact that the greatest radiant heat flux received occurs along a line normal to the radiating surface.

If it is assumed that the lines of heat radiation from a flat panel radiator are not parallel but expanding as shown in Figure 4.8 then the

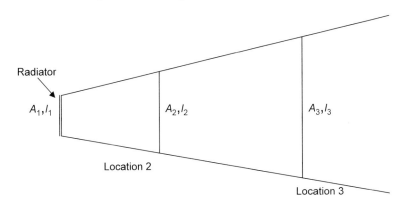

Figure 4.8 Heat radiation received at locations 2 and 3.

total heat radiation received at location 2 will be the same as at location 3.

Thus $I_1 A_1 = I_2 A_2 = I_3 A_3$ W where radiant heat flux I is taken as a mean value at both locations.

The heat flux at each location gets progressively less in direct proportion to the increase in receiving areas.

4.11 Problem solving

The foregoing discussion provides an introduction to the subject of heat radiation.

It should, however, be sufficient to form the underpinning knowledge for the building services engineer. There now follows some examples involving this mode of heat transfer.

Example 4.1
A luminous quartz heater has a temperature of 1200°C and an effective area of 0.3 m × 0.3 m
Determine: (a) the total rate of radiant emission,
(b) the wavelength of maximum energy,
(c) the monochromatic emissive power at the wavelength of maximum energy.

Solution
You will notice that there is no indication of the mean radiant temperature of the surfaces in the enclosure in which the heater is located so the solution will not account for radiation exchange.

Since the heater is luminous it is assumed to be a black body.
(a) Thus from equation (4.2) $I = 5.67 \times 10^{-8}(1200 + 273)^4$
from which

$$I = 2.67 \times 10^5 \text{ W/m}^2 = 267 \text{ kW/m}^2$$

This is the radiant heat flux for the sum of all the wavelengths. The effective area is given as $A = 0.3 \times 0.3 = 0.09 \text{ m}^2$ therefore total radiant emission

$$= 2.67 \times 10^5 \times 0.09 = 2.4 \times 10^4 \text{ W}$$
$$= 24 \text{ kW}$$

(b) From equation (4.5) $\lambda_{max} T = C_3$
thus

$$\lambda_{max} = C_3/T = 2897.6/1473 = 1.97 \text{ μm}$$

(c) From equation (4.3):

$$I = (3.743 \times 10^8 \times 1.97^{-5})/(\exp(1.4387 \times 10^4/1.97 \times 1473) - 1)$$

given that $\exp = e^x = 2.7183^x$ where

$$x = 1.4387 \times 10^4/1.97 \times 1473$$

$$I = (3.743 \times 10^8 \times 1/29.67)/(2.7183^{4.958}) - 1$$
$$I = 1.2615 \times 10^7/(142 - 1)$$
$$I = 8.95 \times 10^4 = 89.5 \, \text{kW/m}^2$$

This is the emissive power at the wavelength of maximum heat radiation.

For the effective area of the radiator the emissive power at this wavelength $= 89.5 \times 0.09 = 8.055 \, \text{kW}$

Summary for Example 4.1
The proportion of the emissive power of $89.5 \, \text{kW/m}^2$ attributed at the wavelength of maximum heat radiation ($1.97 \, \mu\text{m}$), to the total emissive power across all wavelengths of $267 \, \text{kW/m}^2$ is $89.5/267 = 33.5\%$. Thus one-third of the luminous heater output is derived from the wavelength of $1.97 \, \mu\text{m}$.

Example 4.2
Show the effect on radiant heat flux of locating bright aluminium foil having an emissivity of 0.07 in the centre of the cavity of an external cavity wall.

Assume that the two boundaries to the cavity are at $10°\text{C}$ and $1°\text{C}$ respectively, each having an emissivity of 0.9.

Comment upon the effect that the foil will have on the thermal transmittance coefficient for the wall and upon its thermal response.

Solution
Figure 4.9 shows the wall cavity with the foil in place.
Absolute temperatures are used in the solution and:

$$T_1 = (273 + 10) = 283 \, \text{K}, \, T_2 = (273 + 1) = 274 \, \text{K}$$

Equations (4.9) and (4.11) can be adopted here in which T_f is the absolute temperature of the foil. Thus if temperatures remain steady heat flow from T_1 to T_f equals the heat flow from T_f to T_2 and:

$$\sigma(T_1^4 - T_f^4)/((1/e_1) + (1/e_f) - 1) = \sigma(T_f^4 - T_2^4)/((1/e_f) + (1/e_2) - 1)$$

Since the form factor has the same numerical value each side of the heat balance and the Stefan–Boltzmann constant cancels $(T_1^4 - T_f^4) = (T_f^4 - T_2^4)$, and substituting:

$$283^4 - T_f^4 = T_f^4 - 274^4$$
$$283^4 + 274^4 = 2T_f^4$$

from which foil temperature $T_f = 278 \, \text{K}$ $(t_f = 5c)$.

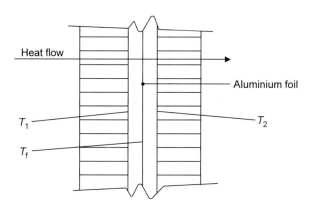

Figure 4.9 External wall cavity with bright aluminium foil in place (Example 4.2).

The radiant heat flux I for two parallel grey surfaces can be determined by combining equations (4.9) and (4.11)

$$I = (F_{1,2})\sigma(T_1^4 - T_2^4) \quad \text{W/m}^2$$

thus:

$$I = \sigma(T_1^4 - T_2^4)/((1/e_1) + (1/e_2) - 1)$$

The radiant heat flux between surfaces 1 and 2 before the foil is located will be:

$$I = 5.67 \times 10^{-8}(283^4 - 274^4)/((1/0.9) + (1/0.9) - 1)$$

and $I = 44.1/1.222 = 36.1 \text{ W/m}^2$

The radiant heat flux when the foil is in position can be determined from either one side or the other of the foil since the heat flux from surface 1 to the foil will equal the heat flux from the foil to surface 2 assuming temperatures remain steady.

Then taking the heat flux from surface 1 to the foil:

$$I = 5.67 \times 10^{-8}(283^4 - 278^4)/((1/0.9) + (1/0.07) - 1)$$

from which $I = 25.03/14.4 = 1.74 \text{ W/m}^2$.

The heat transfer coefficient for radiation $(F_{1,2})h_r$ can be determined from equation (4.15) and for parallel boundaries equation (4.9) is applicable for form factor $F_{1,2}$ and therefore when the foil is in position, working from boundary 1 to the foil:

$$(F_{1,f})h_r = \sigma(T_1 + T_f)(T_1^2 + T_f^2)/((1/e_1) + (1/e_f) - 1)$$
$$(F_{1,f})h_r = 5.67 \times 10^{-8}(283 + 278)(283^2 + 278^2)/((1/0.9) + (1/0.07) - 1)$$
$$(F_{1,f})h_r = 0.348 \text{ W/m}^2\text{K}$$

A much simpler way of finding $(F_{1,f})h_r$ here, since the radiant heat flux I has been calculated, is to use equation (4.13).

Thus $(F_{1,f})h_r = I/dt = 1.74/(283 - 278) = 0.348 \text{ W/m}^2\text{K}$.

The heat transfer coefficient for radiation when the foil is not in place will take place between the two surfaces 1 and 2, thus using again equations (4.9) and (4.15):

$(F_{1,2})h_r = 5.67 \times 10^{-8}(283 + 274)(283^2 + 274^2)/((1/0.9) + (1/0.9) - 1)$

$(F_{1,2})h_r = 4.01 \text{ W/m}^2\text{K}$

Similarly again equation (4.13) can be used here and

$(F_{1,2})h_r = I/dt = 36.1/(283 - 274) = 4.01 \text{ W/m}^2\text{K}$

Summary for Example 4.2

Condition	Radiant heat flux I	Radiant heat transfer coefficient $(F_{1,2})h_r$
no foil	36.1 W/m²	4.01 W/m²K
with foil	1.74	0.348

With the foil in place in the wall cavity the radiant heat flux is reduced to about 5% of its original value. If the foil is 2 mm thick and has a thermal conductivity of 105 W/mK its thermal resistance $R = 0.002/105 = 0.000\,019 \text{ m}^2\text{K/W}$. This will have no significant effect upon the thermal transmittance coefficient (U value) for the wall and therefore no effect upon the heat loss. However, with the foil reflecting back much of the radiant component of heat transfer towards the inner leaf of the wall, the inner leaf will heat up more quickly when the heating plant is started after a shut-down period. The foil will also assist in maintaining the temperature of the inner leaf, and hence indoor temperature, longer when the plant shuts down at the end of the day. This results in energy conservation.

Example 4.3(a)
(i) A vertical panel radiator, fixed to an external wall, measures 1.8 m by 0.75 m high and has a mean surface temperature of 76°C and an emissivity of 0.92.

It is intended to fix bright metal foil having an emissivity of 0.04 directly to the wall behind the radiator having an emissivity of 0.9. The room is held at a mean radiant temperature of 19°C and the wall temperature behind the radiator stabilises at 40°C. Determine the radiant heat emission from the panel before and after the foil is in place.

(ii) If the room air temperature is 21°C determine the heat output by free convection from the radiator. Evaluate the properties at the mean film temperature and assume one of the following relationships for the surface convection coefficient h_c.

For turbulent flow $Nu = 0.1(PrGr)^{0.33}$, for laminar flow $Nu = 0.36(Gr)^{0.25}$

Solution (i)
Figure 4.10 shows the radiator panel fixed to the external wall.

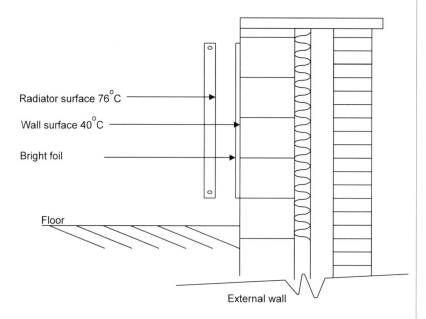

Figure 4.10 Use of bright reflective foil (Example 4.3).

Absolute temperatures are: panel surface, $76 + 273 = 349\,K$
wall, foil surface, $40 + 273 = 313\,K$
room, $19 + 273 = 292\,K$

Radiant heat flow from the back of the panel must account for the form factor for parallel surfaces, equation (4.9).

Radiant heat flow from the front of the panel involves the form factor $F_{1,2} = e_1$. Two separate calculations therefore need to be considered here for each condition.

Considering the back of the panel with no foil in place
From equation (4.9) $F_{1,2} = 1/((1/e_1) + (1/e_2) - 1)$, substituting $e_1 = 0.92$ and $e_2 = 0.9$, $F_{1,2} = 1/1.198 = 0.835$.

From equation (4.11) in which $T_1 = 349\,\text{K}$ and $T_2 = 313\,\text{K}$, $I = 248\,\text{W/m}^2$ and emission $Q = 248 \times 1.8 \times 0.75 = 335\,\text{W}$.

Considering the front of the panel
From equation (4.12) in which

$e_1 = 0.92, T_1 = 349\,\text{K}$ and $T_2 = 292\,\text{K}$,
$I = 395\,\text{W/m}^2$ and emission $Q = 395 \times 1.8 \times 0.75 = 533\,\text{W}$.

Total radiant output from the panel without the foil
 = 335 + 533 = 868 W
You should now confirm these calculations.

Considering the back of the panel with the foil in place
From equation (4.9)

$F_{1,2} = 1/((1/0.92) + (1/0.04) - 1) = 1/25.087 = 0.04$,

from equation (4.11)

$I = 5.67 \times 10^{-8} \times 0.04[(349^4) - (313^4)] = 11.88\,\text{W/m}^2$

and emission $Q = 11.88 \times 1.8 \times 0.75 = 16\,\text{W}$.
The front of the panel will have the same radiant emission as before and $Q = 533\,\text{W}$.
The total radiant output from the panel with the foil in place
 = 533 + 16 = 549 W.

Summary for Example 4.3(i)
With the foil in place the radiant heat output from the panel is 549 W compared with 868 W. The heat loss by radiation through the wall is 868 − 549 = 319 W when the foil is missing. There is therefore a saving in heat energy and hence fuel costs when the foil is used; there will also be a reduction in CO_2 emission into the atmosphere.

Solution to Example 4.3(a)(ii)
From the chapter on heat convection the first step in the solution is to establish whether the free convection over the panel is laminar or turbulent. This is achieved by evaluating the Grashof number

$Gr = \beta g x^3 \rho^2 dt / \mu^2$

The mean film temperature at the panel surface = (76 + 21)/2 = 48.5°C = 321.5 K.
The density and viscosity of the air at the mean film temperature is interpolated from the tables of *Thermodynamic and Transport Properties of Fluids* as $\rho = 1.099\,\text{kg/m}^3$, $\mu = 0.000\,019\,46\,\text{kg/ms}$, $Pr = 0.701$ and $k = 0.027\,89\,\text{W/mK}$.
The coefficient of cubical expansion of the air $\beta = 1/321.5 = 0.003\,11\,\text{K}^{-1}$.
Substituting these values into the Grashof formula:

84   Heat radiation

$$Gr = [(0.00311 \times 9.81 \times (0.75)^3 \times$$
$$(1.099)^2(76 - 21)]/(0.00001946)^2 = 2.27 \times 10^9.$$

Since $Gr > 10^9$ air flow over the panel is turbulent,
thus $Nu = 0.1(0.701 \times 2.27 \times 10^9)^{0.33} = 108.8$
now $Nu = h_c x/k$ from which $h_c = Nuk/x = 108.8 \times 0.02789/0.75 = 4.04 \, W/m^2 K$

Free convective emission $Q = h_c A dt = 4.04 \times (1.8 \times 0.75 \times 2)(76 - 21) = 600 \, W$

Comparison of solutions 4.3(a)(i) with 4.3(a)(ii)
The total output from the panel with the foil in place $= 549 + 600 = 1149 \, W$.
The total output from the panel without the foil $= 868 + 600 = 1468 \, W$.

Example 4.3(b)
A horizontal radiant panel located at high level in a workshop and insulated on its upper face has a surface temperature of 110°C, an emissivity of 0.9 and measures 2.7×1.2 m.

The workshop is held at a mean radiant temperature of 19°C and an air temperature of 15°C.

(i) Determine the total emission downwards.
(ii) Determine the total emission upwards if the outer surface of the insulation has a temperature of 30°C and an emissivity of 0.1.
(iii) Calculate the thickness of thermal insulation given that its thermal conductivity is 0.07 W/mK.

Data: Rational formulae:

$I = (F_{1,2})\sigma[T_1^4 - T_2^4] \, W/m^2$ equation (4.11)

downward $h_c = 0.64((t_s - t_f)/D)^{0.25} \, W/m^2 K$ equation (3.8)

upward $h_c = 1.7(t_s - t_f)^{0.33} \, W/m^2 K$ equation (3.9)

$Q = h_c A dt \, W$ equation (3.16)

Solution (i)
Now $I = 5.67 \times 10^{-8} \times 0.9((383^4) - (292^4))$
from which $I = 727 \, W/m^2$
and $Q_r = 727(2.7 \times 1.2) = 2356 \, W$.
Now $h_c = 0.64((110 - 15)/((2.7 + 1.2)/2))^{0.25}$
from which $h_c = 1.69 \, W/m^2 K$
and $Q_c = 1.69(2.7 \times 1.2)(110 - 15)$,

from which $Q_c = 520$ W.
Total downward emission $Q = 2356 + 520 = 2876$ W

Solution (ii)
The upper surface of the thermal insulation is finished in aluminium foil, hence the low value for its emissivity.
Now $I = 5.67 \times 10^{-8} \times 0.1((303)^4 - (292^4))$
from which $I = 6.57$ W/m^2
and $Q_r = 6.57(2.7 \times 1.2) = 21.3$ W.
Now $h_c = 1.7(30 - 15)^{0.33}$
from which $h_c = 4.155$ W/m^{2}K
and $Q_c = 4.155(2.7 \times 1.2)(30 - 15)$
from which $Q_c = 202$ W.
Total upward emission $Q = 21 + 202 = 223$ W

Solution (iii)
From equation (2.3) $dt_1/R_1 = dt_2/R_2 = I$ W/m^2
Adapting the equation for use here where $t_1 = 110°$C and $t_2 = 30°$C

$$I = (t_1 - t_2)/R_{ins} \text{ W/m}^2$$

substituting: $223/(2.7 \times 1.2) = (110 - 30)/(L/0.07)$
thus: $68.83 = 80/(L/0.07)$.
Rearranging: $68.83 \times L/0.07 = 80$
from which $L = 0.081$ m $= 81$ mm.

Summary for Example 4.3b
Total downward emission $= 2876$ W, total upward emission $= 223$ W, thickness of insulation on the upper side of the panel $= 81$ mm.

Example 4.4
(a) A steam pipe at a temperature of 200°C passes through a room in which the mean radiant temperature is 20°C. A short section of the pipe surface is uninsulated and its emissivity is 0.95. If its area is 0.25 m^2 calculate the rate at which heat radiation will be lost.
(b) How would heat emission be affected by:

(i) painting the pipe with aluminium paint given $e = 0.7$,
(ii) wrapping the pipe tightly with aluminium foil given $e = 0.2$,
(iii) surrounding the pipe with a co-axial cylinder of aluminium foil where its outside diameter is twice the diameter of the steam pipe.

Solution
Absolute temperatures are used and $T_1 = 273 + 200 = 473\,\text{K}$, $T_2 = 273 + 20 = 293\,\text{K}$. (a) Assuming that the uninsulated pipe surface is small compared with the size of the room in which it is located equation (4.12) may be adopted in which $F_{1,2} = e_1$.
Thus

$$I = 5.67 \times 10^{-8} \times 0.95(473^4 - 293^4)$$

and

$$I = 2299\,\text{W/m}^2$$
$$Q = 2299 \times 0.25 = 575\,\text{W}$$

(b) (i) As the only variable is emissivity
$Q = 575 \times \text{ratio of emissivities} = 575 \times 0.7/0.95 = 425\,\text{W}$.
(b) (ii) $Q = 575 \times 0.2/0.95 = 120\,\text{W}$.
(b) (iii) The surface area of the aluminium casing at absolute temperature T_s will be twice that of the pipe. Thus the radiant heat loss from the casing surface to the room will be:

$$Q = 0.2\sigma(T_s^4 - 293^4)(2 \times 0.25).$$

The radiant heat transfer into the aluminium casing from the pipe surface from equations (4.10) and (4.11) will be:

$$I = \sigma(473^4 - T_s^4)/((1/e_1) + A_1/A_2((1/e_2) - 1)\,\text{W/m}^2$$

Accounting for the surface area of the pipe:

$$Q = \sigma(473^4 - T_s^4)(0.25)/((1/e_1) + A_1/A_2(1/e_2) - 1)\,\text{W}$$

substituting data:

$$Q = \sigma(473^4 - T_s^4)(0.25)/((1/0.95) + 0.25/0.5(1/0.2) - 1)$$
$$Q = \sigma(473^4 - T_s^4)(0.25)/3.0526.$$

A heat balance can now be drawn up if temperatures remain steady.
Heat transfer from the pipe to the casing = heat transfer from the casing to the room, thus:

$$\sigma(473^4 - T_s^4)(0.25)/3.0526 = 0.2\sigma(T_s^4 - 293^4)(0.5)$$
$$473^4 - T_s^4 = 1.221(T_s^4 - 293^4)$$
$$5.9054 \times 10^{10} = 2.221 T_s^4$$
from which $T_s = 404\,\text{K}$ ($t_s = 131°\text{C}$)

Thus the radiant heat transfer from the aluminium casing to the room will be:

$$Q = 0.2 \times 5.67 \times 10^{-8}(404^4 - 293^4)(0.5) = 109\,\text{W}.$$

Summary for Example 4.4

Condition of pipe	radiant heat transfer (W)
plain	575
painted with aluminium	425
wrapped in foil	120
encased with aluminium cylinder	109

Example 4.5
A gas-fired radiant heater consumes $5.625\,m^3$ in one and a half hours of natural gas which has a calorific value of $38.4\,MJ/m^3$. The heater has an effective black body temperature of $750°C$ in surroundings at $20°C$. If the area of the heater surface is $0.5\,m^2$ determine the radiant efficiency of the heater.

Solution
From equation (4.2) radiant heat flux $I = \sigma T^4\,W/m^2$
for black body radiation exchange assuming the enclosure is black:

equation (4.8) $I = \sigma(T_1^4 - T_2^4)\,W/m^2$
from which $Q = \sigma(T_1^4 - T_2^4)A_1\,W$
substituting: $Q = 5.67 \times 10^{-8}(1023^4 - 293^4) \times 0.5$
from which radiant output $Q = 31\,kW$

gas input $= (5.625/(3600 \times 1.5)) \times 38\,400$
$\qquad\qquad = 40\,kW$

Radiant efficiency $=$ (output/input) $\times 100$
$\qquad\qquad\qquad = (31/40) \times 100 = 77.5\%$

Example 4.6
A stem thermometer 4 mm in diameter is located in a bend along the axis of a boiler smokepipe 200 mm in diameter. Flue gas temperature is $200°C$ and the reading on the thermometer is $185°C$. The surface temperature of the smokepipe is $140°C$. Determine the convection coefficient for heat transfer between the flue gas and the stem thermometer. Emissivity of the stem thermometer is 0.93 and that of the smokepipe 0.8.

88 Heat radiation

Solution
There is a temperature disparity here between the flue gas, the stem thermometer and the wall of the boiler smokepipe and the reason for it is presented in the summary at the conclusion of the solution. A heat balance may be drawn up as follows.

Convection to the thermometer from the flue gas is equal to the radiation exchange between the thermometer and the wall of the smokepipe.

Absolute temperatures are used and the flue gas is 473 K, the stem thermometer is 458 K and the smokepipe wall is 413 K.

Considering the heat radiation exchange, the form factor for concentric cylinders is found in equation (4.10),

$$1/F_{1,2} = (1/0.93) + (\pi 0.004.L/\pi 0.2L)((1/0.8) - 1)$$

from which $1/F_{1,2} = 1.0803$ and therefore $F_{1,2} = 0.926$

Adapting equation (4.11) $Q = (F_{1,2})\sigma(T_1^4 - T_2^4)A_1$,

substituting the values

$$Q = 0.926 \times 5.67 \times 10^{-8}(458^4 - 413^4)A_1 \text{ W}.$$

Considering the heat transfer by convection, from equation (3.16), $Q = h_c.A.dt$ W and substituting values $Q = h_c A_1(473 - 458)$ W.

Thus combining the formulae into the heat balance

$$h_c A_1(473 - 458) = 0.926 \times 5.67 \times 10^{-8}(458^4 - 413^4)A_1$$

from which $15h_c = 782.69$

and $h_c = 52.18 \text{ W/m}^2\text{K}$.

Summary for Example 4.6
The reason why the stem thermometer, flue gas and smokepipe wall were not at the same temperature was because the boiler smokepipe at 140°C which is 60 K below the flue gas temperature must be rapidly loosing heat to its surroundings. The thermometer would register a truer reading if the smokepipe was adequately insulated. A well-insulated boiler flue pipe will minimize heat transfer between the flue gas, thermometer and smokepipe wall. This analysis applies to other similar applications and accounts for errors in temperature measurement.

4.12 Asymmetric heat radiation

There are three cases of asymmetric heat radiation:

1. Local cooling – radiation exchange with an adjacent cold surface as with a single glazed window.
2. Local heating – radiation exchange with an adjacent hot surface or a series of point sources as with spot lamps.

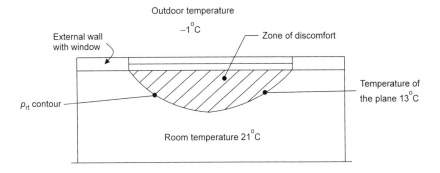

Figure 4.11 Plan of room with an external window showing p_{rt} contour.

3. Intrusion of short wave radiation as with solar radiation through glazing.

Unless all the inner surfaces of an enclosure are at the same temperature mean radiant temperature t_r will vary throughout the space. This variation will produce a change in comfort temperature t_c and introduce asymmetry. Strong asymmetry will promote discomfort.

To quantify the degree of discomfort it is helpful to introduce two concepts:

- plane radiant temperature p_{rt}
- vector radiant temperature v_{rt}

p_{rt} is associated with the effects of radiant cooling which can result, for example, in front of an external single glazed window. Discomfort may result if the p_{rt} when facing the cold window surface is 8 K below the room comfort temperature t_c. Refer to Figure 4.11.

An example of heating discomfort may occur if the v_{rt} resulting from solar irradiation through a glazed window is greater than 10 K above the room comfort temperature. In buildings which are highly intermittent in use such as churches, and in factories where the air temperatures are low, directional high temperature radiant heaters giving a v_{rt} well in excess of 10 K are quite acceptable since they compensate for the low air temperatures and low mean radiant temperatures resulting from the cold enclosing surfaces.

4.13 Solar irradiation and the solar constant

The fraction of the Sun's energy reaching the outer atmosphere of the Earth may be calculated approximately if it is assumed that the Earth travels in a circular path around the Sun. The fraction will be the ratio of the Earth's disc area to that of the spherical surface area described by its radial path around the Sun. The surface area so described will receive all the Sun's radiation whereas the Earth's disc will receive the fraction calculated. Thus the fraction = (area of Earth's disc)/(surface area of described sphere). Refer to Figure 4.12.

Heat radiation

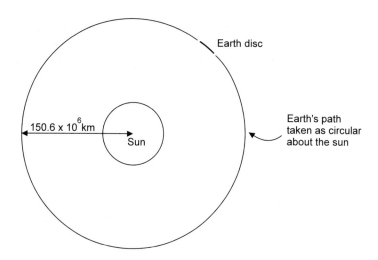

Figure 4.12 Sun and Earth as a disc. The Earth's motion around the Sun considered circular. Fraction of Sun's radiation received by the Earth = $\dfrac{\text{area of Earth's disc}}{\text{surface area of described sphere}}$.

The Earth's radius is 6436 km and the radius of the Earth's path described around the Sun is (150.6×10^6) km,

the fraction $= \pi(6436)^2/4\pi(150.6 \times 10^6) = 4.56 \times 10^{-10}$.

Adapting equation (4.2) the Sun's total emission $Q = \sigma(T^4)A$ W.

The Sun's temperature is 6000 K and its radius is (6985×10^2) km

the total emission $Q = 5.67 \times 10^{-8}(6000)^4 \times 4\pi(6985 \times 10^5)^2$

$$= 4.505 \times 10^{26} \text{ W}$$
$$Q = 4.505 \times 10^{23} \text{ kW}$$

That reaching the Earth $= 4.505 \times 10^{23} \times$ fraction received

$$= 4.505 \times 10^{23} \times 4.56 \times 10^{-10}$$
$$= 2.0543 \times 10^{14} \text{ kW}$$

The solar constant $I = Q/A = 2.0543 \times 10^{14}/\text{disc area kW/m}^2$

$$= 2.0543 \times 10^{14}/\pi(6436 \times 10^3)^2$$
$$I = 1.5786 \text{ kW/m}^2$$

This compares with the measured solar constant perpendicular with the Sun's rays outside the Earth's atmosphere of $I = 1.388$ kW/m^2.

The solar intensity on a horizontal surface at latitude 51.7 on 21 June at 12.00 is: $I = 0.85 \text{ kW/m}^2$.

The solar intensity on a vertical surface facing south at latitude 51.7 on 22 September and 22 March at 12.00 is: $I = 0.700\,\text{kW/m}^2$.

The solar intensities given here are the maximum values in the year for sky clarity of 0.95, cloudiness factor 0.0, ground reflectance factor 0.2 and altitude 0 m to 300 m.

4.14 Solar collectors

There are a variety of solar collectors in use for water heating. Those used for domestic water heating and for heating swimming pools are of the fixed type. Since a typical surface can have a high absorptivity, from Kirchhoff's law, it will also have a high emissivity. A collector will have a selective surface. Refer to Example 4.7 below. If greenhouse type glass is used to protect the collector surface, however, it will allow the passage inwards of short-wave solar radiation and assist in preventing the transmission of long-wave radiation from the collector surface outwards through the glass.

Example 4.7
A flat plate solar collector has a selective surface with an absorptivity of 0.95 and an emissivity of 0.15. The coefficient of convective heat transfer is 3 W/m²K at the collector surface. If the area of the collector is 2 m² and there are four connected to the same system calculate the rate of energy collection and the collection efficiency at a time when the irradiation is 820 W/m². Take the collector temperature as 65°C and outdoor air temperature as 27°C.

Solution
The convection loss from the outer surface of the collector adopting equation (3.16)

$$Q = h_c A \mathrm{d}t = 3 \times (2 \times 4) \times (65 - 27) = 912\,\text{W}$$

The radiation loss

$$Q = \sigma e T^4 = 5.67 \times 10^{-8} \times 0.15 \times (273 + 65)^4 \times (2 \times 4) = 888\,\text{W}$$

The net rate of collection $= (0.95 \times 820 \times 2 \times 4) - 912 - 888 = 4432\,\text{W}$

Solar irradiation $= 820 \times (2 \times 4) = 6560\,\text{W}$

The collection efficiency = net collection/incident irradiation
$= (4432/6560) \times 100 = 67\%$

Note: this is the collection efficiency and not the overall efficiency of the solar collector.

Example 4.8
Calculate the rate of energy absorption on a flat plate collector having an area of 3 m² and positioned normally to the Sun's rays. The surface temperature of the collector is 70°C and outdoor air temperature 21°C. Its absorptivity to solar radiation is 0.95 and the emissivity of the plate is 0.2. The convective heat transfer coefficient at the collector surface is 3 W/m²K. Take the solar constant as 1.388 kW/m² and the transmissivity of the upper atmosphere as 0.63.

Solution
Rate of collection $= 0.95 \times 1388 \times 0.63 = 831$ W/m²
Radiant loss $= 5.67 \times 10^{-8} \times 0.2(273 + 70)^4 = 157$ W/m²
Convection loss $= 3 \times (70 - 21) = 147$ W/m²
Net rate of absorption $= 831 - 157 - 147 = 527$ W/m²
Net rate of collection $= 527 \times 3 = 1581$ W

Summary for Examples 4.7 and 4.8
In both the above examples the net rate of solar irradiation collected is then transferred to the collecting medium which is usually water treated with an antifreeze agent. There is a loss of efficiency here and also at the point where this heated water imparts its energy at the heat exchanger to the water used for consumption. If an efficiency of 60% is taken at each of these points the overall efficiency of the solar collecting system in Example 4.6, and ignoring losses from pipes and storage vessel, will be $0.67 \times 0.6 \times 0.6 = 24\%$.

4.15 Chapter closure

You now have an underpinning knowledge of this mode of heat transfer and have investigated the use of luminous and non-luminous radiant heaters. You understand the importance of the surface characteristics of the radiator and of its location in space for effective results. The concept of asymmetric radiation and discomfort has also been addressed. Some work has been done in this chapter on saving energy by the use of bright aluminium foil. Heat radiation at the wavelength of maximum flux has been shown to represent a significant proportion of the total. You have been introduced to the solar constant and use of solar collectors for water heating.

Historical references

1. Kirchhoff G. Ostwalds Klassiker d. exakten Wissens., 100, Leipzig (1898)
2. Planck M. *The theory of heat radiation* (translation) Dover (1959)
3. Lambert H. L. *Photometria* (1860)
4. Stefan J. Sitzungsber. Akad. Wiss. Wien. Math – naturw. Kl., Vol 79, 391 (1879)
5. Boltzmann L. *Wiedemanns Annalen*, Vol 22, 291 (1884)
6. Hottel H.C. *Notes on radiant heat transmission*. Chem. Eng. Dept. MIT (1951).

Note: Stefan established the constant σ experimentally. Boltzmann subsequently proved it theoretically.

5 Measurement of fluid flow

Nomenclature

a cross-sectional area (m²)
C constant
C_d coefficient of discharge
dh difference in head (m)
dP pressure difference (Pa)
g gravitational acceleration taken as 9.81 m/s² at sea level
h head, metres of fluid flowing (m)
H vertical height (m)
L length of inclined scale (mm)
M mass flow rate (kg/s)
m ratio of cross-sectional areas
p density (kg/m³)
P pressure (Pa)
P_s static pressure (Pa)
P_t total pressure (Pa)
P_u velocity pressure (Pa)
Q volume flow rate (m³/s)
R radius of a circle (m)
S ratio of densities
T absolute temperature (K)
u mean velocity (m/s)
x vertical height (m)
Z height above a datum (m)

5.1 Introduction

This chapter focuses upon the traditional instruments used for measuring gauge pressure, differential pressure and volume flow rate. It also considers the calibration of pressure measuring instruments.

5.2 Flow characteristics

It is helpful to begin with some general definitions relating to the flow of fluids in pipes and ducts.

Uniform flow

The area of cross-section and the mean velocity of the fluid in motion are the same at each successive cross-section.

Example: flow of water through a flooded pipe of uniform bore
Volume flow rate $Q = a \times u \, \text{m}^3/\text{s}$.

Steady flow

The area of cross-section and the mean velocity of the fluid may vary from one cross-section to the next but for each cross-section they do not change with time.

Example: flow of water through a flooded tapering pipe
Volume flow rate $Q = a_1 u_1 = a_2 u_2 \, \text{m}^3/\text{s}$,
thus $u_2 = u_1(a_1/a_2) \, \text{m/s}$.

Continuity of flow

The total amount of fluid entering and leaving a system of pipework or ductwork is the same. This occurs in uniform flow and steady flow.

Example: air flow through a tee piece or junction. See Figure 5.1 in which $Q_1 = Q_2 + Q_3$
Furthermore $u_1 a_1 = (u_2 a_2) + (u_3 a_3)$.

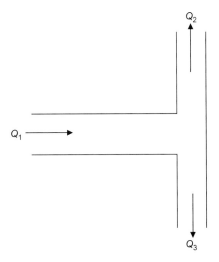

Figure 5.1 Continuity of flow $Q_1 = Q_2 + Q_3$.

Mean velocity

Mean velocity u at any cross-section of area a when the volume flow rate in m³/s is Q will be $u = Q/a$ m/s.

5.3 Conservation of energy in a moving fluid

In order to consider the traditional methods of fluid flow measurement it is necessary to introduce the Bernoulli equation which states that for frictionless flow:

Potential energy Z + Pressure energy $P/\rho g$ + kinetic energy $u^2/2g$ = a Constant.

In this format each energy term in the equation has the units of metres of fluid flowing. Thus for frictionless flow the total energy of the fluid flowing remains constant; no energy is lost or gained in the process.

- Potential energy is that due to a height above a datum.
  Example: water stored in a water tower has potential energy when ground level is taken as datum.
- Pressure energy is that due to static pressure and pump or fan pressure when present.
  Example: water flowing in a heating system subject to the sum of the static head imposed by the feed and expansion tank, and also to the pressure developed by the pump.
- Kinetic energy is that due to the velocity of the fluid in the pipe or duct.

If two points are considered in a system in which fluid is flowing, one downstream of the other as shown in Figure 5.2, the following statement, assuming frictionless flow, can be made:

$$Z_1 + (P_1/\rho g) + (u_1^2/2g) = Z_2 + (P_2/\rho g) + (u_2^2/2g).$$

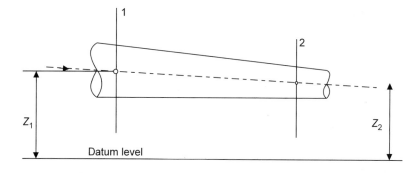

Figure 5.2 Conservation of energy in frictionless flow.

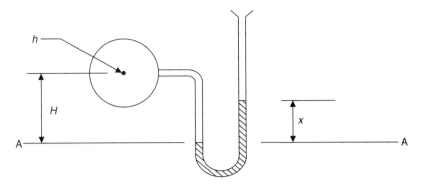

Figure 5.3 The manometer.

Consider Figure 5.3 which shows a manometer open to atmosphere connected to a circular duct. At section A–A the pressures are equal in each limb of the manometer and therefore:

pressure in the left-hand limb = pressure in the right-hand limb

thus $(h + H)\rho_1 g = x\rho_2 g$, gravitational acceleration g cancels and $(h + H) = x\rho_2/\rho_1$;

if $S = \rho_2/\rho_1$
  $h = Sx - H$ m of fluid flowing.

The pressure of the fluid flowing $P = (Sx - H)\rho_1 g$ Pa   (5.1)

5.4 Measurement of gauge pressure with an uncalibrated manometer

Example 5.1
(a) A water-filled manometer is connected to a duct through which air is flowing.
  If the displacement of water levels is 43 mm determine the static pressure generated by the air.
  Data: water density 1000 kg/m³, air density 1.2 kg/m³, and H is 350 mm.
(b) Calibrate the manometer.

Solution (a)
The ratio of densities $S = 1000/1.2 = 833$

substituting into equation (5.1) $P = [(833 \times 0.043) - 0.35] \times 1.2 \times 9.81 = 418$ Pa

Note: since the fluid flowing is air which has a relatively low density H can be ignored without loss of integrity and $P = Sx\rho_1 g$ Pa.

Solution (b)
The displacement of water of 43 mm is equivalent to a static pressure of 418 Pa.

Thus 418/43 = 9.72 Pa per mm of water displacement.

This is a displacement of approximately 1 mm of water for 10 Pa of static pressure and on this basis the manometer can now be calibrated.

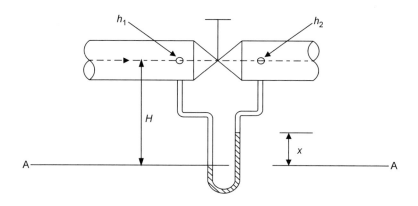

Figure 5.4 The differential manometer.

5.5 Measurement of pressure difference with an uncalibrated differential manometer

Consider Figure 5.4 which shows a differential manometer connected to a pipe transporting fluid. At section A–A the pressures in each limb of the manometer are equal and pressure in the left-hand limb = pressure in the right-hand limb,

thus $(h_1 + H)\rho_1 g = (h_2 + H - x)\rho_1 g + x\rho_2 g$

gravitational acceleration g cancels as does H

and $\qquad (h_1 - h_2)\rho_1 = x\rho_2 - x\rho_1$

therefore $\qquad h_1 - h_2 = x(\rho_2/\rho_1) - x$

then head loss $h_1 - h_2 = (xS - x) = x(S - 1)$ m of fluid flowing

or $\qquad\qquad dh = x(S - 1)$ m of fluid flowing $\qquad\qquad$ (5.2)

therefore pressure loss $dP = x(S - 1)\rho_1 g$ Pa $\qquad\qquad$ (5.3)

Example 5.2
(a) A differential manometer measures the pressure drop across an air filter. The displacement of measuring liquid in the instrument is found to be 10 mm. Determine whether the filter should be replaced.

Data: density of measuring liquid 850 kg/m³, air density 1.2 kg/m³, maximum pressure drop across the filter when it should be replaced 50 Pa.
(b) Calibrate the manometer.

Solution (a)
The ratio of densities $S = 850/1.2 = 708$

substituting the data into equation (5.3) $dP = 0.010(708 - 1) \times 1.2 \times 9.81 = 83.23$ Pa.

Note: unless the ratio of densities S has a relatively low value, equation (5.3) can be reduced to $dP = xS\rho_1 g$ Pa without loss of integrity.

Clearly the filter is in need of replacement since an excess pressure drop across it implies that it is partially clogged.

Solution (b)
The displacement of liquid of 10 mm results in a pressure drop of 83.23 Pa.

Thus $83.23/10 = 8.23$ Pa per mm of displacement.

This is a displacement of approximately 1 mm for 8 Pa of differential pressure and on this basis the instrument can now be calibrated.

Inclined differential monometers

The displacement of the measuring fluid has been considered in a limb of the manometer which is in the vertical position (Figure 5.4). The calibration can be difficult to read accurately because of the small scale. Incline manometers are used to ensure more accurate readings.

If the angle of inclination of the measuring limb is changed from the vertical to 20° from the horizontal, Figure 5.5, then cosine $70 = 1/L$ from which $L = 1/\text{cosine}$ and $70 = 1/0.342 = 2.92$. Thus the scale length is now 2.9 times that of the corresponding vertical scale. Alter-

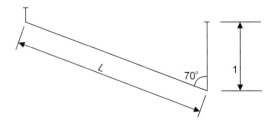

Figure 5.5 The inclined manometer, $L = 1/\cos 70$.

natively, 1 mm displacement on the vertical scale equals 2.9 mm on the scale inclined at 20° to the horizontal.

In the case of the solution to Example 5.2, the displacement of water of 10 mm on the vertical scale of the differential manometer would now be extended on the inclined scale to $10 \times 2.9 = 29$ mm and the calibration would now be $83.23/29 = 2.87$ Pa per mm of displacement. You should now confirm that this is so.

5.6 Measurement of flow rate using a venturi meter and orifice plate

The venturi and orifice plate are instruments specifically made for each application. Once installed they are permanently fixed in position.

Consider Figure 5.6 which shows a venturi fitted in a horizontal pipe in which fluid is flowing. The venturi is a fixed instrument and purpose made for the application.

The design of the venturi meter requires that the entry or converging cone has an angle of 21°, the length of the throat is equal to its diameter and the diverging cone has an angle of 5° to 7°. The two tappings measure static pressure and may be bosses or piezometer rings.

Applying the Bernoulli equation for frictionless flow at sections 1 and 2:

$$Z_1 + (P_1/\rho g) + (u_1^2/2g) = Z_2 + (P_2/\rho g) + (u_2^2/2g),$$

since the pipe is horizontal $Z_1 = Z_2$,

rearranging the equation $(P_1 - P_2)/\rho g = (u_2^2 - u_1^2)/2g$.

We know the equation rearranges in this way since $u_2 > u_1$ thus $P_1 > P_2$, the equation may now be written as $dh = (u_2^2 - u_1^2)/2g$

since $dh = dP/\rho g$.

For continuity of flow $a_1 u_1 = a_2 u_2$ in which d_1 and d_2 are fixed for the chosen application.

So $u_2 = (a_1/a_2) \times u_1$.

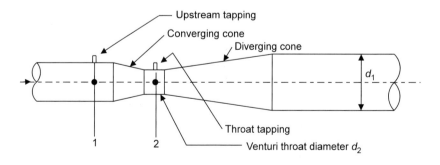

Figure 5.6 The venturi meter.

If area ratio $m = a_1/a_2 = d_1^2/d_2^2$, $u_2 = mu_1$

substituting $dh = [(mu_1)^2 - u_1^2]/2g$

then $dh = (u_1^2/2g)(m^2 - 1)$ m of fluid flowing.

In units of pressure $dP = (u_1^2/2g)(m^2 - 1)\rho g$ Pa

$dP = (\rho u_1^2/2)(m^2 - 1)$ Pa

and rearranging $u_1 = [(2dP)/\rho(m^2 - 1)]^{0.5}$ m/s

therefore $Q_1 = u_1 a_1 = [(2dP)/\rho(m^2 - 1)]^{0.5} \times a_1$

the formula is rearranged thus $Q_1 = [2/\rho(m^2 - 1)]^{0.5} \times a_1 \times (dP)^{0.5}$

where $[2/\rho(m^2 - 1)]^{0.5} \times a_1 = C$, a constant for the instrument and based upon its physical dimensions for the chosen application and the density of the fluid flowing.

Thus $Q = C(dP)^{0.5}$ m³/s

This formula is derived from the Bernoulli equation for frictionless flow. Clearly there will be a small loss due to friction as fluid passes through it. The coefficient of discharge C_d, determined for each instrument, accounts for this and for the venturi meter it varies between 0.96 and 0.98. It is found empirically before leaving the manufacturer.

Therefore actual flow $Q = C_d C(dP)^{0.5}$ m³/s (5.4)

The orifice plate is shown in Figure 5.7. This also is a fixed instrument designed for a specific application. The determination of flow rate is

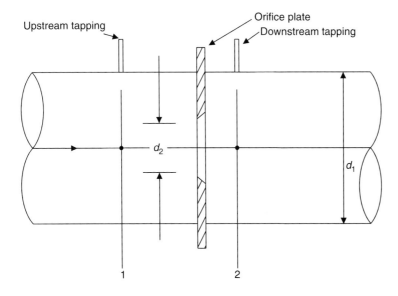

Figure 5.7 The orifice plate.

obtained from the same equation as that for the venturi meter. The coefficient of discharge for the orifice plate C_d varies between 0.6 and 0.7.

Venturi meters are used for measuring the flow rate of liquids. The orifice plate is normally used for measuring the flow rate of gases such as steam since it has less effect upon the compressibility of the substance.

Example 5.3
A venturi meter is located in a horizontal pipeline transporting water at 75°C and is connected to an uncalibrated differential manometer whose calibration limb is inclined at 20° to the horizontal.

From the data find the constant C for the meter.

If the inclined manometer shows a displacement of measuring liquid of 70 mm determine the volume flow and hence the mass flow rate of water in the pipe.

Data: pipe diameter 50 mm, venturi throat diameter 25 mm, water density 975 kg/m³, density of measuring liquid is 13 600 kg/m³ and the coefficient of discharge for the meter C_d is 0.96.

Solution
The manufacturer of the venturi would have requested details of the temperature of the water flowing, the design flow rate and the diameter of the pipe. The constant C would then have been supplied with the instrument. Here we are asked to calculate it.

It was established that constant $C = [2/\rho(m^2 - 1)]^{0.5} \times a_1$
where $m = d_1^2/d_2^2 = (50/25)^2 = 4$
substituting: $C = [2/975(16 - 1)]^{0.5} \times (\pi \times 0.05^2/4) = 0.000\,023$.
Actual flow rate through the meter $Q = C_d C (dP)^{0.5}$ from equation (5.4).

A differential manometer already calibrated in Pa or mbar would be used to measure the pressure drop across the tappings on the venturi meter. Here we have to calculate it. The differential pressure loss dP for the water flowing is determined from equation (5.3).

However the vertical displacement x is required and since cosine $70 = x/L$, where L = the calibration limb at 20° to the horizontal, $x = L\cos 70$, from which $x = 70 \times 0.342 = 24$ mm.

From equation (5.3), $dP = x(S - 1)\rho g$ where $S = 13\,600/975 = 13.95$
thus $dP = 0.024(13.95 - 1) \times 975 \times 9.81 = 2973$ Pa
substituting into equation (5.4)
$Q = 0.96 \times 0.000\,023 \times (2973)^{0.5} = 0.0012$ m³/s

or $Q = 1.2 \, \text{l/s}$
and mass flow $M = Q\rho = 0.0012 \times 975 = 1.174 \, \text{kg/s}$.

Summary for Example 5.3
The calibration for this differential manometer measuring water flowing at 75°C would be $2973/70 = 42.5 \, \text{Pa}$ for each millimetre displacement of measuring fluid.

Notice the effect of temperature on the density of the fluid flowing and therefore the differential pressure reading. If the water flowing was at 5°C its density would be $1000 \, \text{kg/m}^3$ and pressure loss $dP = 0.024(13.6 - 1) \times 1000 \times 9.81 = 2967 \, \text{Pa}$, constant $C = [2/1000(16 - 1)]^{0.5} \times (\pi \times 0.05^2/4) = 0.000\,022\,7$
flow rate $Q = 0.96 \times 0.000\,022\,7(2967)^{0.5} = 0.001\,187 \, \text{m}^3/\text{s}$ and mass flow $M = Q\rho = 0.001\,187 \times 1000 = 1.187 \, \text{kg/s}$.

Example 5.4
An orifice plate is installed in a steam main for measuring the flow of steam. Determine the rate of flow given the manufacturer's coefficient of discharge as 0.7 and the manufacturer's constant for the instrument as $C = 0.0046$.

The differential pressure measured at the orifice plate was 290 mbar. Given the steam density as $3.666 \, \text{kg/m}^3$ find the mass flow rate of steam in the pipe.

Solution
The manufacturer's constant C derives from $C = [2/\rho(m^2 - 1)]^{0.5} \times a_1$. The manufacturer must therefore be provided with the steam pressure and quality in order to establish its density, and the pipe diameter into which the orifice plate is to be fitted as a permanent device.

The measured pressure drop of 290 mbar $= 29\,000 \, \text{Pa}$.

From equation (5.4) $Q = C_d C(dP)^{0.5} \, \text{m}^3/\text{s}$, substituting, the volume flow rate of steam $Q = 0.7 \times 0.0046 \times (29000)^{0.5} = 0.548 \, \text{m}^3/\text{s}$.
The mass transfer of steam $M = 0.548 \times 3.666 = 2 \, \text{kg/s}$.

Example 5.5
A venturi meter is fitted into a horizontal water main and is intended to act as a means of recording the water flow rate to a process.

To do this the pressure tappings are connected to the ends of a cylinder of 20 mm bore fitted with a piston which has a pen

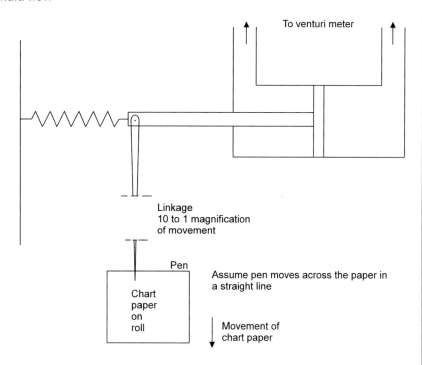

Figure 5.8 Use of venturi meter and chart recorder, (Example 5.5).

connected to the piston rod by means of a linkage in such a way that each millimetre of rod movement causes the pen to move 10 mm across the paper.

The system is shown diagrammatically in Figure 5.8. The rate of water flow may vary between 240 and 170 l/s during the process operation. The water main is 300 mm bore and the throat of the venturi is 200 mm bore.

(a) Determine the velocity of water through the pipe and the force on the piston at each of the two flow rates. Ignore the diameter of the piston rod. Take water density as 1000 kg/m^3.
(b) If the spring extends 4 mm per Newton, determine the minimum width of chart paper needed to record the flow rates between the two limits.

Solution (a)
Since $Q = ua, u = Q/a = 4Q/\pi d^2$ and for a flow rate of 0.24 m^3/s, $u_1 = 0.24 \times 4/\pi \times (0.3)^2 = 3.395$ m/s and $u_2 = 0.24 \times 4/\pi \times (0.2)^2 = 7.639$ m/s, similarly for a flow rate of 0.17 m^3/s, $u_1 = 2.405$ m/s and $u_2 = 5.411$ m/s.

Adopting the Bernoulli equation in section 5.3 and taking section 1 at the tapping on the upstream pipe and section 2 at the tapping on the throat of the venturi:

$$Z_1 + (P_1/\rho g) + (u_1^2/2g) = Z_2 + (P_2/\rho g) + (u_2^2/2g).$$

Since the venturi is horizontal $Z_1 = Z_2$.

Since $d_1 > d_2, u_2 > u_1$ and rearranging the equation:

$$(P_1 - P_2)/\rho g = (u_2^2 - u_1^2)/2g.$$

For a flow rate of $0.24 \, m^3/s$

$$(P_1 - P_2)/\rho g = [(7.639)^2 - (3.395)^2]/2g \text{ and } dP = 23\,417 \, Pa.$$

Now $dP =$ force/area,
so force $= dP \times$ area of piston $= 23\,417 \times \pi(0.02)^2/4 = 7.357 \, N$,
similarly for a flow rate of $0.17 \, m^3/s$, dP is calculated as $11\,749 \, Pa$.
You should now confirm this calculation.
Using the above result, force $= 11\,749 \times \pi(0.02)^2/4 = 3.69 \, N$.

Solution (b)

Spring movement $= (7.357 - 3.69) \times 4 = 14.7 \, mm$.

Linkage $= 14.7 \times 10 = 147 \, mm$.

Therefore chart width should be a minimum of 150 mm.

The inclined venturi

Normally the venturi meter is positioned horizontally. It is, however, pertinent to consider the effect upon a reading when the instrument is located in an inclined position. Figure 5.9 shows such a location.

At section A–A, the head exerted in the left-hand limb = the head exerted in the right-hand limb.

Thus $P_1/\rho g + Z_1 = P_2/\rho g + (Z_2 - x) + S_x$ where S is the ratio of densities of the measuring fluid and the fluid flowing.

Rearranging: $[(P_1 - P_2)/\rho g] + Z_1 - Z_2 = S_x - x = x(S - 1) = dh$.

This is equation (5.2) for head loss in a differential manometer.

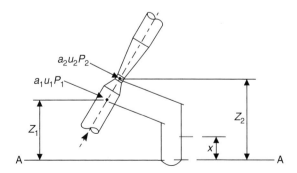

Figure 5.9 The inclined venturi meter.

Thus the final equation (5.4) is independent of Z_1 and Z_2 and there is no effect on the measurements taken from a venturi meter located in the inclined position.

5.7 Measurement of air flow using a pitot static tube

The pitot static tube, unlike the venturi and orifice plate, is a portable instrument and used to measure air flow. It consists of two tubes in a coaxial arrangement as shown in Figure 5.10. The inner tube connected to the nose of the instrument measures total pressure in the air stream; the outer tube has holes in its sides which measure the static pressure of the air stream. By connecting each tube to a differential manometer the velocity pressure of the air stream is obtained.

Consider the Bernoulli theorem for the total energy at a point in a system of air flow $(Z + P/\rho g + u^2/2g)$ in metres of air flowing = total energy in the air. Z, the potential energy due to the height above a datum is insignificant because air density is relatively very low. If energy is measured in units of pressure:

$$(P/\rho g)\rho g + (u^2/2g)\rho g = \text{total energy of the moving air in Pa},$$

thus $P + (\rho/2)u^2 = $ total energy.

The first term, P is the static pressure energy generated by the fan working on the air and the second term is the pressure energy due to the velocity of the moving air. These terms are commonly called static pressure P_s and velocity pressure P_u. Thus total pressure of the moving air at a point $P_t = P_s + P_u$ Pa where $P_u = (\rho/2)u^2$ Pa.

When the pitot tube is connected to a differential manometer velocity pressure is obtained and $P_u = P_t - P_s$. Figure 5.11 shows the static and total pressure tubes of the pitot tube separated to identify the equivalent readings.

Rearranging the formula for velocity pressure P_u, mean air velocity

$$u = (2P_u/\rho)^{0.5} \text{m/s} \tag{5.5}$$

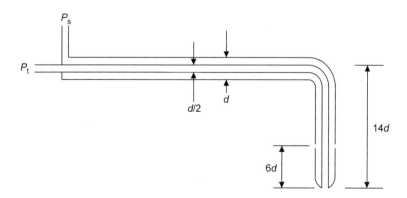

Figure 5.10 The pitot static tube.

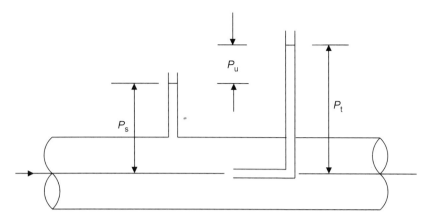

Figure 5.11 The equivalent readings from a pitot static tube.

This is the theoretical velocity. Actual velocity = $C(2P_u/\rho)^{0.5}$ m/s and actual flow rate $Q = C(2P_u/\rho)^{0.5} \times a$ m³/s.

For the pitot static tube constant $C = 1.0$ for Reynolds numbers greater than 3000 and where the cross-sectional area of the pitot tube is insignificant compared with the cross-sectional area of the duct a.

Thus when air flow is in the turbulent region

$$Q = (2P_u/\rho)^{0.5} \times a \text{ m}^3/\text{s} \tag{5.6}$$

The accuracy of readings of velocity pressure depends very much on the person using the pitot tube ensuring that the nose of the instrument is pointing at the air stream and parallel to the duct. Readings are best taken on a straight section of duct where the velocity profile of the air is more likely to be regular.

Air velocity distribution

For square and rectangular ducts the cross-section is divided into a grid so that the velocity pressure, velocity and hence volume flow Q is the sum of the flow rates in each segment of the grid, thus:

$$Q = a_1 u_1 + a_2 u_2 + a_3 u_3 + \ldots = \Sigma(au) \text{ m}^3/\text{s}$$

and mean velocity $= \Sigma(au)/\Sigma a$ m/s.

For circular ducts the velocity profile on straight sections is the same across any diameter, therefore for an annular ring of mean radius r and area a

$$\text{volume flow } Q = \Sigma(u_r a_r) \text{ m}^3/\text{s}$$

and mean velocity $= \Sigma(u_r a_r)/\pi R^2$ m/s.

Air density varies with temperature and pressure. Standard air density ρ_1 at 20°C and 101 325 Pa is 1.2 kg/m³ and for other tempera-

tures and pressures air density ρ_2 can be found by applying the gas laws:
thus

$$\rho_2 = \rho_1(T_1/T_2)(P_2/P_1) \, \text{kg/m}^3 \qquad (5.7)$$

Air density can also be found from the tables of *Thermodynamic and Transport Properties of Fluids* at constant standard air pressure and different absolute temperatures.

Example 5.6
Air is conveyed in a duct of section 300×450 mm. A series of readings of velocity pressure are obtained with the aid of a pitot static tube and calibrated manometer. The mean value is determined as 65 Pa. Determine the mean air velocity in the duct and hence the volume flow rate.
  Data: temperature of the air flowing $35°C$, barometric pressure 748 mm mercury, density of mercury $13\,600\,\text{kg/m}^3$.

Solution
Local atmospheric pressure
  $P = h\rho g = 0.748 \times 13\,600 \times 9.81 = 99\,795\,\text{Pa}$
absolute temperatures
  $T_1 = 273 + 20 = 293\,\text{K}, T_2 = 273 + 35 = 308\,\text{K}$
atmospheric pressures
  $P_2 = 99\,795\,\text{Pa}, P_1 = 101\,325\,\text{Pa}$
from equation (5.7)
  $\rho_2 = 1.2(293/308)(99\,795/101\,325) = 1.124\,\text{kg/m}^3$
from equation (5.5)
  $u = (2P_u/\rho)^{0.5} = (2 \times 65/1.124)^{0.5} = 10.75\,\text{m/s}$
from equation (5.6)
  $Q = 10.75 \times (0.3 \times 0.45) = 1.45\,\text{m}^3/\text{s}.$

5.8 Chapter closure

You are now able to determine gauge pressure, differential pressure and flow rate using the traditional pressure measuring instruments and flow measuring devices.

The working limits of the pitot static tube have been discussed. You can also calibrate the pressure measuring instruments described in this chapter and the inclined manometer has been introduced to obtain more accurate readings.

This work can now be extended to other examples and problems in the field of measurement of pressure, differential pressure and flow rate.

Characteristics of laminar and turbulent flow 6

Nomenclature

A	area of cross-section (m²)
d	characteristic dimension (m)
d	diameter m (mm)
dP	pressure drop (Pa)
dp	specific pressure drop (Pa/m)
dx	unit length (m)
f	frictional coefficient in turbulent flow
g	gravitational acceleration (m/s²)
ks	absolute roughness (mm), surface roughness
L	characteristic dimension (m)
L	length (m)
M	mass flow rate (kg/s)
ρ	density (kg/m³)
Pw	power (W)
Q	volume flow rate (m³/s)
R	radius (m)
r	radius (m)
μ	absolute viscosity (kg/ms)
γ	kinematic viscosity (m²/s)
x	characteristic dimension (m)

6.1 Introduction

Fluid viscosity is the measure of the internal resistance sustained in a fluid being transported in a pipe or duct as one layer moves in relation to adjacent layers. At ambient temperature heavy fuel oils, for example, possess a high viscosity while the lighter oils possess a low viscosity. The walls of the pipe or duct provide the solid boundaries for the fluid flowing and because of the friction generated between the boundary and the fluid interface, which has a drag effect, and fluid viscosity, the velocity of flow varies across the enclosing boundaries to produce a velocity gradient. In a straight pipe or duct maximum fluid velocity would be expected to occur along the centre-line and zero velocity at the boundary surfaces.

Characteristics of laminar and turbulent flow

There are therefore two factors to consider when water, for example, flows along a pipe, namely the viscosity of the water and the coefficient of friction at the pipe inside surface.

Fluid viscosity is temperature dependent and the coefficient of friction at the inside surface of the pipe or duct is velocity dependent as well as being related to surface roughness and a characteristic dimension of the pipe, namely pipe diameter.

This sets the scene for a discussion on laminar and turbulent flow.

6.2 Laminar flow

In about 1840 a Frenchman by the name of Poiseuille and an American by the name of Hagen identified the following equation, which is dedicated to them, during experiments on fluid viscosity:

$$Q = (\pi r^4 dP)/(8\mu dx) \, m^3/s \qquad (6.1)$$

This formula can be rearranged in terms of pressure drop per metre run of pipe or duct dp

thus: $(dP/dx) = dp = 8Q\mu/\pi r^4$

since $Q = uA, u = Q/A = Q/\pi r^2$

substituting for Q $dp = 8\mu u/r^2 = 8\mu u/(d/2)^2$

thus $\qquad dp = 32\mu u/d^2 \, Pa/m \qquad (6.2)$

and $\qquad dP = 32\mu u x/d^2 \, Pa \qquad (6.3)$

This equation can be expressed in terms of head loss dh of fluid flowing and since

$dP = dh\rho g$ Pa, and substituting this for dP

$dh = 32\mu u x/\rho g d^2$ m of fluid flowing $\qquad (6.4)$

This arrangement of the Poiseuille/Hagen formula is probably better known than equation (6.1). You will notice that the effect of friction between the fluid and the boundary interface is not accounted for. This is because the dominating feature resisting fluid flow is the viscosity of the fluid. Laminar flow is therefore sometimes referred to as viscous flow. It is not too clear whether Hagen or Poiseuille fully understood the characteristics of laminar flow during the process of establishing their formula which shows that head loss dh is proportional to the ratio of u/d^2. This is verified by their reaction to the claim by another Frenchman called Darcy, who in about 1857 proposed a different formula in which head loss due to friction is proportional to the ratio of u^2/d. It was left to an Englishman by the name of Osborne Reynolds to reconcile the dispute in 1883 at his famous presentation to the Royal Society. Refer to Figure 6.1.

Reynolds established that Darcy on the one hand and Hagen/Poiseuille on the other were both correct and that the different formulae were the result of different types of fluid flow. He found that in

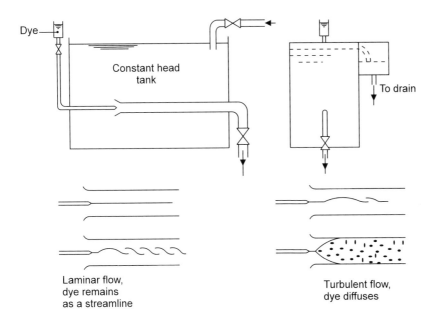

Figure 6.1 Osborne Reynolds' experiment (1883).

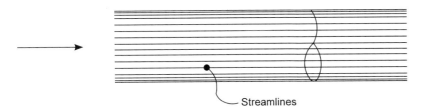

Figure 6.2 Streamlines in laminar flow.

laminar flow the fluid moved along streamlines which are parallel to the pipe wall (Figure 6.2). Any disturbance in a straight pipe in which the fluid was moving in laminar flow would cause a disturbance along the streamline which would dissipate at some point downstream and return to a streamline. The experiment which Reynolds presented to the meeting of the Royal Society used coloured dye to illustrate the phenomenon. He was able to show that as fluid velocity increased the streamline could not be maintained, and the point was reached when the dye suddenly diffused in the water which was used for the experiment and flow became turbulent. Turbulent flow therefore can be identified as the random movement of fluid particles in a pipe or duct with the sum of the movements being in one direction. Refer to Figure 6.4.

112 Characteristics of laminar and turbulent flow

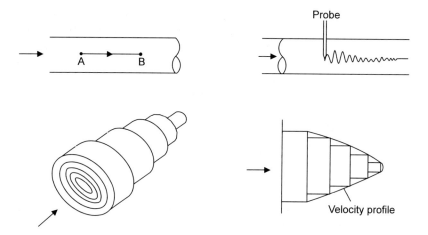

Figure 6.3 Laminar flow.

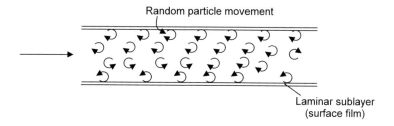

Figure 6.4 Fluid in turbulent flow.

Laminar flow characteristics, Figure 6.3

- A particle at point A at some time will be at point B after travelling in a straight line parallel to the tube.
- A disturbance in the fluid generated by the insertion of a probe will straighten out downstream.
- The change in velocity of the fluid across the pipe section is not linear but parabolic.
- It is convenient to think of the fluid motion as a series of concentric layers slipping over one another, and the distances by which each layer is extruded represents the velocity of each layer.

6.3 Turbulent flow

As fluid velocity is increased in the pipe or duct the flow changes from laminar to turbulent. This is known as the critical point. If the velocity of fluid through a pipe in which flow is known to be laminar is increased slowly until flow just becomes turbulent the higher critical point is reached.

If now the velocity is slowly decreased, flow will at first remain turbulent then at a velocity lower than that at which turbulence commenced flow will again become laminar. This is known as the lower critical point. Between the higher critical point and the lower critical point fluid flow is unstable. The Frenchman Darcy had identified a formula for turbulent flow in 1857 which, before the work done by Reynolds, could not be reconciled with the formula of Poiseuille and Hagen. The Darcy equation is:

$$dh = 4fLu^2/2gd \quad \text{m of fluid flowing} \quad (6.5)$$

In the Darcy formula fluid viscosity μ is not present but frictional coefficient f is. Here the dominant feature is the frictional resistance to flow at the boundary surface, and the effects of the fluid viscosity in terms of resistance to flow are insignificant. Turbulent flow for this reason is sometimes called frictional flow.

Osborne Reynolds found that a dimensionless group of variables could be used to reconcile equations (6.4) and (6.5) and his name is used to identify the dimensionless group as the Reynolds number Re.

$$Re = \rho u d/\mu \quad (6.6)$$

It can also be expressed as

$$Re = dM/\mu A = ud/\nu \quad (6.7)$$

Reynolds' experiments identified the following general guidelines.

For pipes and circular ducts when $Re < 2000$ flow is said to be in the laminar region and equation (6.4) can be adopted. Between an Re of 2000 and 3500 flow is in transition and therefore unstable. Above an Re of 10 000 flow is said to be turbulent and the Darcy equation (6.5) may be used.

For practical calculations equation (6.5) is also used for Reynolds numbers in excess of 3500.

The Darcy equation for turbulent flow can be expressed in terms of volume flow rate in a similar way to the Poiseuille/Hagen equation (6.1)

$$dh = 4fLu^2/2gd \text{ m of fluid flowing}$$
since $Q = uA, u = Q/A = 4Q/\pi d^2$,

substituting for u in the Darcy equation:

$$dh = (4fL(4Q/\pi d^2)^2)/2gd = (64/2g\pi^2)(fLQ^2/d^5) = (1/3)fLQ^2/d^5$$

from which

$$Q = (3dhd^5/fL)^{0.5} \text{ m}^3/\text{s} \quad (6.8)$$

This is known as Box's formula.

Turbulent flow characteristics, Figure 6.4

- When the fluid velocity is high disturbances in the fluid are not damped out.
- Fluid particles as well as travelling along the pipe also travel across it in a random manner.
- Fluid particles cannot pass through the pipe wall and as the pipe surface is approached these perpendicular movements must die out. Thus turbulent flow cannot exist immediately in contact with the solid boundary.
- Even when the mean velocity is high resulting in a high Re number and the greater part of the boundary layer is turbulent there remains a very thin layer adjacent to the solid boundary in which flow is laminar. This is called the laminar sublayer.

SUMMARY

Most if not all fluid flow in the context of building services is in the turbulent region although it is well to check that this is so from the Reynolds number before proceeding with the solution to a problem.

6.4 Boundary layer theory

The velocity of a mass of fluid in motion which is subject to gravity and is remote from solid boundaries is uniform and streamline. There is no velocity gradient and hence there is no shear stress in the fluid. The viscosity of the fluid is therefore not affecting fluid motion and neither is the friction at the fluid boundary interface. Fluid in contact with a solid boundary is brought to rest. Further away fluid will be slowed but by not as much as that closer to the boundary. Thus near solid boundary surfaces the effects of friction and fluid viscosity result in a velocity gradient. Refer to Figure 6.5. For fluid flow in flooded pipes or ducts, the pipe/duct walls act as the solid boundary where

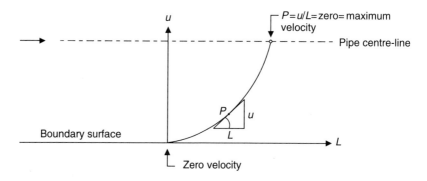

Figure 6.5 Velocity gradient – fluid flow in circular conduits.

fluid velocity is said to be zero as a result of fluid viscosity and the frictional resistance at the fluid/boundary interface.

L. Prandtl, who is considered the founder of fluid mechanics, defined the theory of the boundary layer in a variety of applications, at the turn of the century in Hanover and subsequently in Gottingen where he founded the Kaiser Wilhelm Institute.

The building services engineer is mainly concerned with a limited number of applications such as the flow of air in ducts and the flow of water in pipes and channels. If a flat plate is positioned in a stream of flowing fluid which is unaffected by solid boundaries, the development of the boundary layer from the leading edge of the plate can be identified, one side being considered. Refer to Figure 6.6.

The following points can then be observed:

- fluid velocity under the boundary layer starts at zero at the leading edge of the plate and reaches a maximum at the boundary limit;
- the thickness of the boundary layer is very small compared with its length L;
- there are three discrete regions;
- laminar and transition lengths are very short; therefore flow is often considered turbulent throughout the whole boundary layer;
- during transition Re has critical values;
- the plate imposes a resistance to flow causing a loss in fluid momentum. The plate experiences a corresponding force called skin friction;
- the boundary layer increases in thickness to a maximum value as the length L from the leading edge of the plate increases;
- at points close to the solid boundary of the flat plate velocity gradients are large and the viscous shear mechanism is significant enough to transmit the shear stress to the boundary, such that the layer adjacent to the boundary is in laminar motion even when the rest of the boundary layer is turbulent. This is the laminar sublayer which you will notice becomes extremely thin downstream of the leading edge of the flat plate;

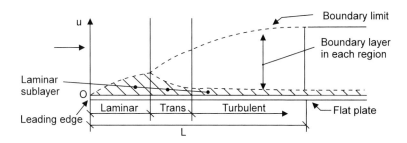

Figure 6.6 Formation of the boundary layer on one side of a flat plate.

116　Characteristics of laminar and turbulent flow

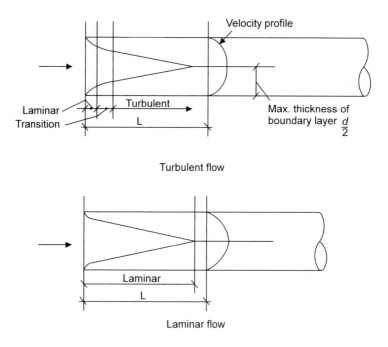

Figure 6.7 Formation of the boundary layer in a pipe.

- a pipe may be considered as a flat plate wrapped round to reform itself. Thus fluid velocity starts at zero at the pipe wall and reaches a maximum value at the centre-line of a straight pipe. It then returns to zero velocity at the opposite wall of the pipe forming the velocity profile which is bullet shaped. The length L from the leading edge of the flat plate becomes infinite when the plate is reformed into a pipe since the thickness of the boundary layer is restricted at the pipe centre-line by the boundary layer from the pipe wall opposite. If the pipe or duct was very large in diameter this may not be so. The length L therefore for most practical applications becomes the straight length of the pipe or duct being considered.

Figure 6.7 shows the formation of the boundary layer in laminar and turbulent flow in a straight pipe (see also Figure 6.15).

Velocity profile for laminar and turbulent flow in straight pipes

Due to the surface resistance at the boundary walls of the pipe and the viscosity of the fluid, maximum velocity occurs at the pipe centre-line and zero velocity at the pipe wall. The velocity gradient u/L may be obtained at any point P on the velocity profile, Figure 6.5. At the pipe centre-line the velocity profile $u/L = 0$. Thus the boundary layer is the layer of fluid contained in a velocity profile up to the point where the velocity gradient is zero.

Boundary layer separation

The separation of the boundary layer from the solid boundary surface does not occur in straight pipes or ducts. This is because there is a steady static pressure loss in the direction of flow. It does occur however in tees, Y junctions, bends and gradual enlargements and its effects on pressure losses through fittings are analysed in Chapter 7. It can be shown that in each of the fittings identified here there is a momentary gain in static pressure as the fluid passes through. This is most commonly noted in the gradual enlargement in which the gain in static pressure is held. The gain in static pressure is at the expense of a corresponding loss in velocity pressure whether it is momentary or otherwise and this causes the boundary layer to separate from the solid boundary surface. It rejoins at some point downstream. Figures 6.8, 6.9, 6.10 and 6.11 illustrate the phenomenon.

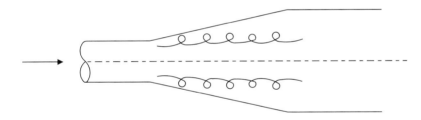

Figure 6.8 Boundary separation in an enlargement.

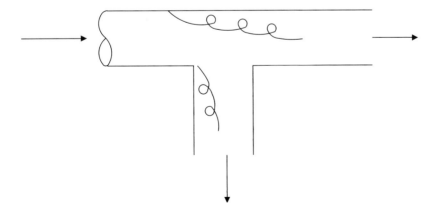

Figure 6.9 Boundary separation in a tee piece.

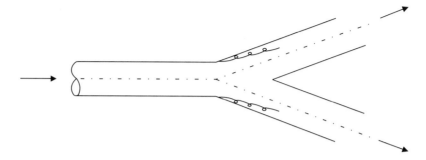

Figure 6.10 Boundary separation in a Y junction.

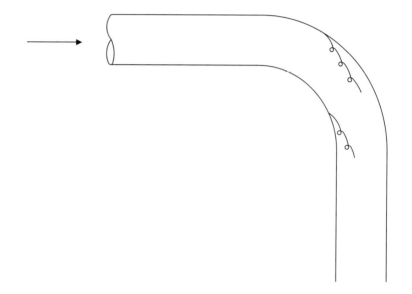

Figure 6.11 Boundary separation in a 90° bend.

6.5 Characteristics of the straight pipe or duct

The coefficient of friction f appears in the Darcy equation (6.5) but does not figure in equation (6.4) of Poiseuille/Hagen. The reason for the omission is because the roughness of the pipe wall is not a significant factor in laminar flow.

The coefficient of friction f at the fluid boundary is a function of a lineal measurement of the high points on the rough internal surface ks called surface roughness/absolute roughness and measured in millimetres. It is also a function of a characteristic dimension of the pipe, taken as its diameter d, or in the case of a rectangular section the shorter side, also measured in millimetres.

The coefficient of friction therefore is dependent upon the relative roughness which is the ratio of absolute roughness and the internal pipe diameter ks/d. Table 6.1 lists the surface roughness factors ks for various materials.

Table 6.1 Surface roughness factors for conduits. (Reproduced from *CIBSE Guide* section C4 (1986) by permission of the Chartered Institution of Building Services Engineers.)

Material	ks in mm
Non-ferrous drawn tubing including plastics	0.0015
Black steel pipe	0.046
Aluminium ducting	0.05
Galvanized steel piping and ducting	0.15
Cast-iron pipe	0.20
Cement or plaster duct	0.25
Fair faced brick or concrete ducting	1.3
Rough brickwork ducting	5.0

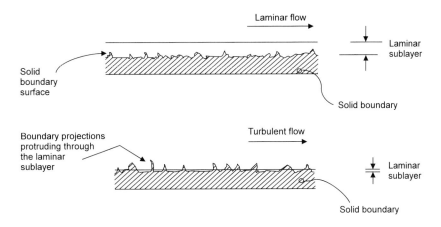

Figure 6.12 Surface roughness and the laminar sublayer (surface film).

Even with turbulent flow the effect of fluid viscosity and friction at the boundary surface results in a film at the boundary wall which is known as the laminar sublayer. Under certain conditions this may be sufficiently thick to obscure the high points on the boundary surface and flow will be as for a smooth pipe. The film thickness reduces with increasing velocity and at some high value of Re rough projections protrude, increasing turbulence (Figure 6.12) which in heat exchangers, for example, assists heat transfer.

6.6 Determination of the frictional coefficient in turbulent flow

A formula has been developed by Colebrook and White for the resolution of the frictional coefficient f in the Darcy equation (6.5)

$$1/(f)^{0.5} = -4\log((ks/3.7d) + 1.255/(Re)(f)^{0.5})$$

120 Characteristics of laminar and turbulent flow

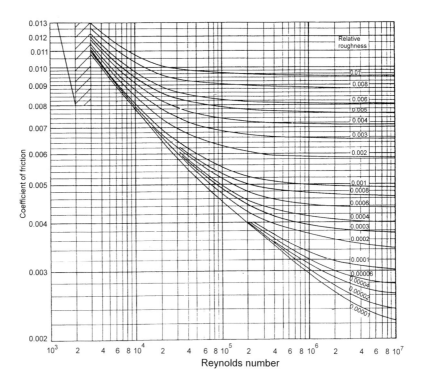

Figure 6.13 The Moody chart for turbulent flow. (Reproduced from the *CIBSE Guide* (1986) by permission of the Chartered Institution of Building Services Engineers.)

It can be seen that a simple solution to evaluate f using this formula requires a process of iteration. An alternative method of solution involves the use of the Moody chart of Poiseuille and Colebrook–White. Figure 6.13 shows the Moody chart taken from section C of the *CIBSE Guide*. You will see that after evaluating the Reynolds number and the relative roughness ks/d for a particular application the coefficient of friction f can be obtained by reading off the left-hand axis of the chart.

6.7 Solving problems

A number of problems and their solutions relating to laminar and turbulent flow are included in Chapter 7 which introduces and applies the theorem for the conservation of energy first proposed by Daniel Bernoulli in 1738. The problems considered here specifically relate to the two types of fluid flow.

You will see that fluids with low values of absolute viscosity have high values for the Reynolds number at relatively low mean velocities. It is important to remember that absolute viscosity varies with fluid temperature. For air it increases with increase in temperature, but for water it decreases with temperature rise. Reference should be made to the tables of the *Thermodynamic and Transport Properties of Fluids* by Rogers and Mayhew.

Example 6.1
A horizontal galvanized steel pipe is 80 mm nominal bore and 50 m in length. Determine the pressure loss sustained along the pipe if cold water flows at a mean velocity of 1.5 m/s. Determine also the specific pressure loss. Take water density as 1000 kg/m^3 and absolute viscosity as 0.001 306 kg/ms.

Solution
From equation (6.6) Reynolds number $Re = \rho u d/\mu = 1000 \times 1.5 \times 0.08/0.001\,306 = 91\,884$, flow is therefore turbulent.

From Table 6.1 $ks = 0.15$ and relative roughness $= ks/d = 0.15/80 = 0.001\,875$.

By locating the Reynolds number and the relative roughness on the Moody chart, Figure 6.13, the coefficient of friction $f = 0.0063$.

Adopting the Darcy equation for turbulent flow equation (6.5)

$dh = 4fLu^2/2gd = 4 \times 0.0063 \times 50 \times (1.5)^2/2 \times 9.81 \times 0.08$

$\quad = 1.806$ m of water flowing

$dP = dh \times \rho \times g = 1.806 \times 1000 \times 9.81 = 17\,719$ Pa.

Summary for Example 6.1
1. The specific pressure loss $dp = dP/L = 17\,719/50 = 354$ Pa/m. Specific pressure loss in pipe sizing is regulated by the maximum mean water velocity to avoid the generation of noise. For steel pipes above 50 mm nominal bore this is 3 m/s or 4 m/s in long straight runs.
2. Referring to the pipe sizing tables in the *CIBSE Guide* Section C, and given a water velocity of 1.5 m/s and a calculated specific pressure drop of 354 Pa/m in 80 mm galvanized pipe at 10°C, the mass flow rate is interpolated as 7.22 kg/s.
   Using the data in Example 6.1:

the volume flow rate of water $\quad Q = u \times \pi \times d^2/4 = 1.5 \times \pi \times (0.08)^2/4$
$\quad\quad\quad\quad\quad\quad\quad\quad = 0.00\,754\,\text{m}^3/\text{s}$

$Q = 7.54\,\text{l/s}$

and for cold water where $\rho = 1000\,\text{kg/m}^3$

$M = 7.54\,\text{kg/s}$

This is close to the interpolated reading from the pipe sizing table of $M = 7.22$ kg/s.

3. It is helpful to find the maximum mean water velocity attainable in laminar flow here. This occurs when $Re < 2000$ and using equation (6.6):

$Re = \rho u d/\mu$ then $u = Re\mu/\rho d = 2000 \times 0.001\,306/1000 \times 0.08$
$= 0.033$ m/s.

This velocity is really too low for economic pipe sizing and hydraulic regulation. In fact laminar flow rarely exists in systems of water distribution in building services.

Example 6.2
Oil is pumped through a straight pipe 150 mm nominal bore and 80 m long. It discharges 10 m above the pump and neglecting all losses other than friction, determine:

(a) the power required to pump 16.67 kg/s of oil along the pipeline.
(b) the maximum flow rate of oil that the pipe can transport in laminar flow and the pump power required.

Take oil density as 835 kg/m³ and viscosity as 0.12 kg/ms.

Solution (a)
From equation (6.6) $Re = \rho u d/\mu$ where mean velocity

$u = M/\rho A = 4M/\rho \pi d^2$ m/s.

Have a look at the units of the equation for u:

$(kg/s)/(kg/m^3)\,(m^2) = m/s$

thus $u = 4 \times 16.67/835 \times \pi \times (0.15)^2 = 1.13$ m/s.
Substituting into the Reynolds formula

$Re = 835 \times 1.13 \times 0.15/0.12 = 1179$,

thus $Re < 2000$ and oil flow is laminar.
Adopting Poiseuille's equation (6.4) $dh = 32\mu u x/\rho g d^2$ m of oil flowing, substituting

$dh = 32 \times 0.12 \times 1.13 \times 80/835 \times 9.81 \times (0.15)^2 = 1.88$ m
Pump head required dh = viscous loss + elevation

$dh = 1.88 + 10 = 11.88$ m of oil flowing

Pump power $P_w = Mgdh$ W

Having a look at the units of the terms in the equation for P_w:

$(kg/s)(m/s^2)(m) = (kg\,m/s^2)(m/s)$

where $(kg\,m/s^2)$ are the basic SI units for force in N since Force = mass × acceleration.

Thus the units for power $P_w = N(m/s) = Nm/s = W$
therefore substituting for pump power $P_w = 16.67 \times 9.81 \times 11.88$
$$= 1943 \, W.$$

Solution (b)
Laminar flow exists up to a maximum Reynolds number Re of 2000.

From equation (6.6) $Re = \rho u d / \mu$ thus maximum mean velocity
$u = \mu Re / \rho d$
substituting:
$$u = 0.12 \times 2000 / 835 \times 0.15 = 1.916 \, m/s$$
since it was found in part (a) that $u = M/\rho A$, then $M = uA\rho \, kg/s$
then
$$M = 1.916 \times (\pi \times (0.15)^2 / 4) \times 835 = 28.274 \, kg/s$$
from equation (6.4)
$dh = 32 \mu u x / \rho g d^2 = 32 \times 0.12 \times 1.916 \times 80 / 835 \times 9.81 \times (0.15)^2$
$dh = 3.19 \, m$ of oil flowing.

Pump head required dh = viscous loss + elevation
$$dh = 3.19 + 10 = 13.19 \, m \text{ of oil flowing.}$$
Pump power
$$P_w = Mgdh = 28.274 \times 9.81 \times 13.19 = 3658 \, W.$$

Summary for Example 6.2

Mass flow M	pipe diameter d	Reynolds number Re	mean velocity u	pump power P_w
16.67 kg/s	150 mm	1179	1.13 m/s	1943 W
28.274	150	2000	1.916	3658 W

Conclusions to Example 6.2
The pump power is the output power and does not account for pump efficiency.

An increase in mass flow of $(28.274 - 16.67)/16.67 = 70\%$ results in an increase in pump power required of $(3658 - 1943)/1943 = 88\%$.

Fluid viscosity is temperature dependent. The viscosity of fuel oils is particularly sensitive to temperature and medium and heavy fuel oils require heating before pumping can begin. The pipeline will also need to be well insulated and, depending upon its viscosity, may require tracing to maintain the temperature and hence satisfactory oil flow.

Laminar flow as well as turbulent flow can occur in systems of oil distribution.

Example 6.3
Given that the velocity at radius r for laminar flow is expressed as:

$$u = dP(R^2 - r^2)/4\mu L$$

where dP is the pressure drop over length L and R is the pipe inside radius, show that the maximum velocity is twice the mean velocity.

Solution
From Poiseuille's equation (6.4) $dh = 32\mu uL/\rho gd^2$ m of oil flowing. Rearranging in terms of mean velocity u:

$$u = dh\rho gd^2/32\mu L \text{ m/s}$$

since $dP = dh\rho g$, mean velocity $u = dPd^2/32\mu L$ m/s
Actual velocity at radius r: $u = dP(R^2 - r^2)/4\mu L$ m/s
Maximum velocity will occur at the pipe centreline when $r = 0$.
Thus maximum velocity: $u = dPR^2/4\mu L$.
Since maximum velocity = twice the mean velocity

$$dPR^2/4\mu L = 2(dPd^2/32\mu L)$$

then since $R^2 = (d/2)^2$, $dP(d/2)^2/4\mu L = dPd^2/16\mu L$
therefore

$$dPd^2/16\mu L = dPd^2/16\mu L$$

thus maximum velocity = twice the mean velocity in laminar flow.

Example 6.4
Two cold-water tanks each of 4500 litres capacity are refilled every two hours.
  The vertical height of the water main is 26 m and its horizontal distance from the water utility's main is 9 m. If the available pressure is 300 kPa during peak demand calculate the diameter of the rising main.
  Data: pressure required at the ball valve is 30 kPa, make an allowance for pipe fittings of 10% on straight pipe, assume initially that the coefficient of friction f is 0.007, the viscosity of cold water is 0.001 306 kg/ms and the water density is 1000 kg/m³.

Solution
Figure 6.14 shows the arrangement in elevation.
  Flow rate required in the rising main = $4500 \times 2/2 \times 3600 = 1.25$ l/s.

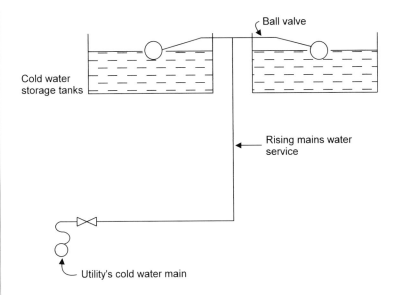

Figure 6.14 Example 6.4.

Mains pressure available for pipe sizing

$= 300 -$ static lift pressure $-$ pressure at ball valve

$= 300 - h\rho g - 30$

$= 300 - (26 \times 1000 \times 9.81/1000) - 30$

$dP = 300 - 255 - 30 = 15\,\text{kPa}$

since $dP = dh\rho g$, $dh = dP/\rho g = 15\,000/1000 \times 9.81 = 1.53\,\text{m}$ of water. Total equivalent length of pipe and fittings $= (26 + 9)1.1 = 38.5\,\text{m}$.

Initially assuming turbulent flow and adopting Box's formula equation (6.8)

$Q = (3dhd^5/fL)^{0.5}$ and rearranging in terms of pipe diameter d:

$d = (fLQ^2/3dh)^{1/5}$

Substituting:
$d = (0.007 \times 38.5 \times (0.00\,125)^{2/3} \times 1.53) = (9.1742 \times 10^{-8})^{1/5}$

$d = 0.039\,\text{m}$

thus standard pipe diameter $d = 40\,\text{mm}$.

It is now necessary to check that water flow is turbulent and to verify the value of the coefficient of friction f.
Since $Q = uA = u\pi d^2/4$,
mean velocity $u = 4Q/\pi d^2 = 4 \times 0.00\,125/\pi(0.04)^2 = 1\,\text{m/s}$
from equation (6.6) $Re = \rho u d/\mu = 1000 \times 1 \times 0.04/0.001\,306$
$= 30\,628$
thus flow is turbulent and adopting Darcy's equation (6.5)

$dh = 4fLu^2/2gd$.

Rearranging in terms of frictional coefficient $f = 2dhgd/4Lu^2$, substituting:

$f = 2 \times 1.53 \times 9.81 \times 0.04/4 \times 38.5 \times 1^2$

$f = 0.0078$

The frictional coefficient used in the solution was $f = 0.007$. The effect on the pipe diameter can be shown by recalculation where $d = (fLQ^2/3dh)^{1/5}$ and substituting using $f = 0.0078$ this time,

$d = (0.0078 \times 38.5 \times (0.00125)^2/3 \times 1.53)^{1/5}$

pipe diameter $d = 0.04\,\text{m} = 40\,\text{mm}$.

Clearly the small error in the initial value for the coefficient of friction is insignificant here.

Summary for Example 6.4

The solution to this problem can be achieved by applying Bernoulli's theorem for the conservation of energy. This is discussed in Chapter 7. If the theorem is applied here taking section 1 to be at incoming mains level and section 2 to be at tank level:

$Z_1 + (P_1/\rho g) + (u_1^2/2g) = Z_2 + (P_2/\rho g) + (u_2^2/2g) + \text{losses}$

since $u_1 = u_2$

$((P_1 - P_2)/\rho g) + Z_1 - Z_2 = \text{losses}$

substituting: $((300\,000 - 30\,000)/\rho g) + 0 - 26 = fLQ^2/3d^5$
thus

$1.53 = 0.007 \times 38.5 \times (0.00125)^2/3d^5$

from which pipe diameter d can be evaluated and $d = 40\,\text{mm}$. You should now confirm this solution.

Example 6.5
Air at 27°C flows at a mean velocity of 5 m/s in a 30 m straight length of galvanized sheet steel duct 400 mm diameter. Determine the static pressure loss along the duct due to air flow.

Data taken from the tables of *Thermodynamic and Transport of Fluids* for dry air at 300 K:

$\mu = 0.00001846\,\text{kg/ms},\ \rho = 1.177\,\text{kg/m}^3$.

Solution
The type of flow can be identified from the Reynolds number, equation (6.6):

$Re = \rho u d/\mu = 1.177 \times 5 \times 0.4/0.00001846 = 127\,935$

since $Re > 2000$ flow is turbulent and Darcy's equation (6.5) can be used to find the pressure loss in the duct. However, it is first of all necessary to find the coefficient of friction f in the Darcy equation and this can be done by using the Moody chart, Figure 6.13. From Table 6.1 the absolute roughness of the duct wall is 0.5 mm and relative roughness $= k_s/d = 0.15/400 = 0.000\,375$.

Using the calculated value of Re and relative roughness the coefficient of friction, from the Moody chart is $f = 0.0047$.

Substituting into the Darcy equation

$dh = 4 \times 0.0047 \times 30 \times 5^2/2 \times 9.81 \times 0.4$

$dh = 1.8$ m of air flowing

since $dP = dh\rho g$ pressure loss $dP = 1.8 \times 1.177 \times 9.81 = 20.8$ Pa.

Summary for Example 6.5
1. The specific pressure drop $dp = dP/L = 20.8/30 = 0.7$ Pa/m. A typical rate of pressure drop in straight ducts for low pressure ventilation and air conditioning systems in which the maximum mean air velocity is around 8 m/s is around 1.0 Pa/m.
2. Two pressures are present in a system of air flow namely static pressure and velocity pressure. This is discussed more fully in Chapters 5 and 7.
3. It is assumed for the purposes of duct sizing that air behaves as an incompressible fluid. This is not necessarily the case at the prime mover or fan where its operating characteristic can show the effects of compression. You should refer to fan manufacturers' literature here.
4. It is helpful to consider the maximum mean air velocity attainable in laminar flow in the duct. For laminar flow $Re < 2000$ thus using equation (6.6)

   $2000 = 1.177 \times u \times 0.4/0.00001846$

from which $u = 0.08$ m/s
This velocity is too low for duct-work design; air flow is therefore invariably in the turbulent region.

6.8 Chapter closure

This chapter has provided you with the underpinning knowledge of the two models of fluid flow, namely laminar and turbulent, relating to the flow of water, oil and air in pipes and ducts. It has given you a methodology for identifying the type of flow in a system and procedures for solving some problems. The text also defines the pressure losses and mean fluid velocities which one may find in systems conveying oil, air and water. The characteristics of laminar and turbulent

128 Characteristics of laminar and turbulent flow

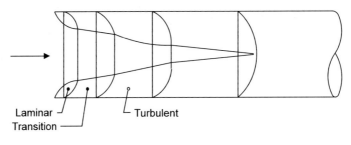

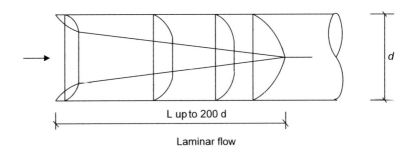

Figure 6.15 Development of velocity profiles – flow in pipes.

flow are discussed in the context of containment in the solid boundaries of the pipe or duct.

Further work is investigated in Chapters 5 and 7. Chapter 7 includes partial flow in pipes and flow in open channels. Part of Chapter 11 focuses on the dimensionless numbers used here and elsewhere in the book.

Flow of fluids in pipes, ducts and channels 7

Nomenclature

a	area of cross-section (m^2)
A	surface area (m^2)
b	breadth (mm) (m)
C	Bernoulli constant, Chezy constant
C_c	coefficient of contraction
d	diameter, depth (mm) (m)
dh	head loss in m of fluid flowing
dP	pressure loss (Pa) (kPa)
dp	specific pressure loss (Pa/m)
E	energy Nm (J)
f	coefficient of friction
g	gravitational acceleration at sea level (9.81 m/s^2)
i	hydraulic gradient (m/m)
k	velocity head/pressure loss factor
K	constant
k_s	surface roughness (mm), absolute roughness
L	length (m)
m	mass (kg)
m	hydraulic mean depth/diameter (m)
M	mass transfer (kg/s)
n	roughness coefficient
P	pressure (Pa) (kPa)
P	permeability
Q	volume flow rate (m^3/s) (l/s)
R	rainfall intensity (mm/h)
Re	Reynolds number
S	ratio of densities
u	mean velocity of flow (m/s)
x	displacement of measuring fluid (mm) (m)
ρ	density (kg/m^3)
Z	vertical height in relation to a datum (m)
μ	viscosity (kg/ms) (Ns/m^2)

Flow of fluids in pipes, ducts and channels

7.1 Introduction

This chapter focuses upon the determination of mass transfer of fluids subject to a prime mover and to gravity. This forms a significant part of the design of heating, ventilating and air conditioning systems and hot and cold water supply. It also focuses on pressure loss resulting from frictional flow and on the hydraulic gradient.

7.2 Solutions to problems in frictionless flow

In Chapter 5 Bernoulli's conservation of energy at a point for a moving fluid or a stationary fluid having potential energy, was introduced and stated that:

$$\left.\begin{array}{r}\text{potential energy plus pressure} \\ \text{energy plus kinetic energy}\end{array}\right\} = \text{total energy} = \text{a constant}$$

thus $Z + (P/\rho g) + (u^2/2g) = C$ in metres of fluid flowing.

In pressure units each of the terms in the Bernoulli theorem must be multiplied by ρ and g

thus $(Z\rho g) + P + (\rho u^2/2) = C\,\text{Pa}$

In energy units of joules or Nm the Bernoulli theorem in metres of fluid flowing must be multiplied by m and g,

thus $(Zmg) + (Pm/\rho) + (mu^2/2) = C\,\text{J or Nm}$

The dimensions of these terms can be checked to ensure integrity. This process is considered in detail in Chapter 10.

Chapter 5 also introduced the Bernoulli statement that the total energy of a moving fluid at one point in a system is equal to the total energy of that fluid at some point downstream. The following example illustrates this statement.

Example 7.1

The suction pipe of a pump rises from a ground storage tank at a slope of 1 in 7 and cold water is conveyed at 1.8 m/s.

If dissolved air is released when the pressure in the pipe falls to more than 50 kPa below atmospheric pressure, find the maximum practicable length of pipe ignoring the effects of friction. Assume the water in the tank is at rest.

Solution

Figure 7.1 shows a diagram of the system in elevation. Applying the Bernoulli equation for the conservation of energy at points 1 and 2 in the system and taking atmospheric pressure as 101 325 Pa:

$$Z_1 + (P_1/\rho g) + (u_1^2/2g) = Z_2 + (P_2/\rho g) + (u_2^2/2g)$$

thus by substitution

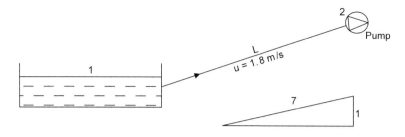

Figure 7.1 Example 7.1.

$$Z_1 + (101\,325/\rho g) + 0 = Z_2 + [(101\,325 - 50\,000)/\rho g] + (1.8)^2/2g$$

rearranging

$$Z_2 - Z_1 = [(101\,325 - 51\,325)/\rho g] - (1.8)^2/2g$$
$$dZ = 5.097 - 0.165 = 4.932\,\text{m}$$

This is the vertical length of the allowable rise from the ground storage tank. Since the pump is located at the upper point it is called suction lift. The maximum length of suction pipe for a gradient of 1 in 7 is

$$L = 7 \times 4.932$$
$$L = 34.5\,\text{m}$$

Summary for Example 7.1
As the gradient increases so the maximum practicable length decreases and for a gradient of 1 in 3, practical length $L = 14.8\,\text{m}$. You should confirm that this is so. And for a vertical pipe, practical length $L = 4.93\,\text{m}$.

These practical lengths relate to a subatmospheric pressure in the suction pipe between sections 1 and 2 of 50 kPa, hence the term suction lift. The theoretical maximum vertical length (suction lift) for cold water in this pipe will occur when atmospheric pressure in the suction pipe at section 2 is zero. It can be obtained by applying the Bernoulli equation, thus:

$$Z_1 + (P_1/\rho g) + (u_1^2/2g) = Z_2 + (P_2/\rho g) + (u_2^2/2g)$$

substituting: $Z_1 + (101\,325/1000g) + 0 = Z_2 + 0 + (1.8)^2/2g$
rearranging: $Z_2 - Z_1 = 10.33 - 0.165 = 10.164\,\text{m}$
and maximum theoretical vertical lift $L = 10.16\,\text{m}$.

This assumes a mean water velocity in the suction pipe of 1.8 m/s.
This amount of suction lift is impossible to achieve. At zero atmospheric pressure within the suction pipe, water vaporizes at

0°C and it will therefore evaporate before reaching the impeller of the pump which anyway is not designed to handle vapour. As the pump generates negative pressure in the suction pipe, water will be drawn up it to a point where its absolute pressure corresponds to its saturation temperature and partial evaporation occurs. Priming the pump will not assist it to achieve a suction lift of this magnitude.

The maximum practical suction lift for cold water is about 5 m. Since the water pressure in the pipe is subatmospheric the pipe must not be made from collapsible material such as canvas. If water must be pumped from a point lower than 5 m, a submersible pump is employed and located in the water contained in the tank or well.

Example 7.2
A 75 mm bore siphon pipe rises 1.8 m from the surface of water in a tank and drops to a point 3.6 m below the water level where it discharges water to atmosphere. Ignoring the effects of friction, determine the discharge rate in l/s and the absolute pressure of the water at the crest of the siphon. Take atmospheric pressure as equivalent to 10 m of water and water density as 1000 kg/m³.

Solution
Figure 7.2 shows the arrangement in elevation.
Adopting the Bernoulli equation for frictionless flow at points 1 and 3 taking point 3 as datum:

$$Z_1 + (P_1/\rho g) + (u_1^2/2g) = Z_3 + (P_3/\rho g) + (u_3^2/2g)$$

where $Z_1 = 3.6$ m, $Z_3 = 0$, $u_1 = 0$, $P_1 = P_3 = 10$ m
substituting: $3.6 + 10 + 0 = 0 + 10 + (u_3^2/2g)$

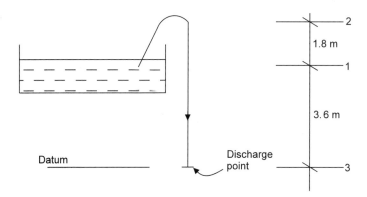

Figure 7.2 Example 7.2.

from which $u_3^2 = 3.6 \times 2g = 70.632$
and $u_3 = 8.4 \, \text{m/s}$

Rate of discharge:
$$Q = u \times \pi d^2/4 = 8.4 \times \pi \times 0.075^2/4 = 0.037 \, \text{m}^3/\text{s}$$
$$Q = 37 \, \text{l/s}$$

Equating points 1 and 2 and keeping point 3 as datum:
$$Z_1 + (P_1/\rho g) + (u_1^2/2g) = Z_2 + (P_2/\rho g) + (u_2^2/2g)$$
for uniform flow $u_1 = u_3 = 8.4 \, \text{m/s}$
substituting: $3.6 + 10 + 0 = (3.6 + 1.8) + (P_2/\rho g) + 8.4^2/2g$
rearranging $P_2 = [(13.6 - 5.4) - (8.4^2/2g)]\rho g$
from which $P_2 = (13.6 - 5.4 - 3.6)1000g = 45\,126 \, \text{Pa}$

Thus the absolute water pressure at the crest of the siphon = 45 kPa.

Alternatively sub-atmospheric pressure at this point is 45 kPa.

From the summary to Example 7.1 you can see that the effect of the low-water pressure at the crest of the siphon may cause the water to separate. It is likely therefore that the discharge will be erratic.

Example 7.3
A 65 mm bore fire hydrant is fed from a water tank located 37 m vertically above. A pressure gauge and stop valve are fitted at the hydrant and with the valve fully open water flows at 26 l/s. Determine the gauge pressure reading:

(a) with the hydrant valve fully open;
(b) with the valve shut.

Solution (a)
Figure 7.3 shows the system in elevation.
Adopting the Bernoulli equation for the conservation of energy at points 1 and 2
$$Z_1 + (P_1/\rho g) + (u_1^2/2g) = Z_2 + (P_2/\rho g) + (u_2^2/2g)$$
where water velocity in the tank $u_1 = 0$. Water velocity in the pipe
$$u_2 = Q/a = 0.026 \times 4/\pi d^2 = 0.026 \times 4/\pi \times (0.065)^2$$
$$= 7.835 \, \text{m/s}.$$

Pressure P_1 at the tank is atmospheric and therefore the gauge pressure is zero.

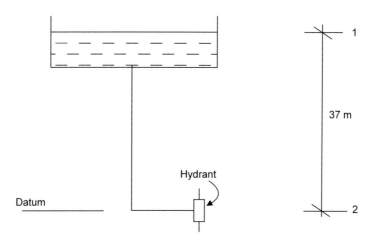

Figure 7.3 Example 7.3.

substituting:
$$37 + 0 + 0 = 0 + P_2/\rho g + [(7.835)^2/2g]$$
from which
$$P_2 = (37 - 3.13) \times 1000 \times 9.81 = 332\,226\,\text{Pa}$$
and
$$P_2 = 332\,\text{kPa}$$

Solution (b)
With no flow $u_1 = u_2 = 0$
substituting in the Bernoulli equation:
$$37 + 0 + 0 = 0 + (P_2/\rho g) + 0$$
from which
$$P_2 = 37 \times 1000 \times 9.81 = 362\,970\,\text{Pa}$$
and
$$P_2 = 363\,\text{kPa}$$

Example 7.4
Air at 30°C flows in a horizontal circular duct and at section A mean velocity is 7 m/s and the static pressure is 300 Pa. If the duct expands gradually from 380 mm at point A find the duct diameter downstream at point B where the static pressure is registered as 320 Pa. Take standard air density at 20°C as 1.2 kg/m³.

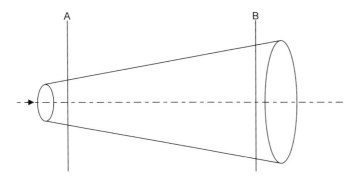

Figure 7.4 Example 7.4.

Solution
Figure 7.4 shows the system in elevation. Adopting the Bernoulli equation for frictionless flow:

$$Z_a + (P_a/\rho g) + (u_a^2/2g) = Z_b + (P_b/\rho g) + (u_b^2/2g)$$

As the duct is horizontal $Z_a = Z_b$.
Ignoring the effect of pressure variations on air density:

for air at 30°C $\rho_2 = \rho_1(T_1/T_2) = 1.2(273+20)/(273+30) = 1.16 \text{ kg/m}^3$.

Thus $0 + (300/1.16 \times 9.81) + (7^2/2g) = 0 + (320/1.16 \times 9.81) + (u_b^2/2g)$

and $\qquad 26.36 + 2.5 = 28.12 + (u_b^2/2g)$
from which $\qquad u_b^2 = 0.74 \times 2 \times 9.81 = 14.5$
so $\qquad u_b = 3.81 \text{ m/s}$

Volume flow rate $Q_a = u_a \times \pi d_a^2/4 = 7 \times \pi \times 0.38^2/4$
$\qquad\qquad\qquad\quad = 0.794 \text{ m}^3/\text{s}$

For steady flow $\quad Q_a = Q_b$
thus $\qquad\qquad 0.794 = 3.81 \times \pi \times d_b^2/4$
from which $\qquad d_b^2 = 4 \times 0.794/3.81 \times \pi = 0.265$
and $\qquad\qquad d_b = 0.515 \text{ m} = 515 \text{ mm diameter}$.

Summary for Example 7.4
You will have seen that as the duct transformation piece is an expansion, mean air velocity decreases and static pressure increases. This is known as static regain. Refer to Example 7.13.

7.3 Frictional flow in flooded pipes and ducts

Consider the following example as an introduction to frictional flow.

Example 7.5
A jet of water issuing at a velocity of 22.5 m/s is discharged through a fire hydrant nozzle having a diameter of 75 mm.

(a) Determine the power of the issuing jet if the nozzle is supplied from a reservoir 30 m vertically above.
(b) What is the loss of head in the pipeline and nozzle?
(c) What is the efficiency of power transmission?

Take the density of water as 1000 kg/m³.

Solution (a)
The energy at the nozzle $= Z + (P/\rho g) + (u^2/2g)$ m of water flowing.

At the nozzle the potential energy is zero and the pressure energy is converted to kinetic energy, thus:

energy at the nozzle $= 0 + 0 + u^2/2g = 22.5^2/2g = 25.8$ m of water.

Power $= Q dP$

where $Q = ua = 22.5 \times \pi(0.075^2)/4 = 0.0994$ m³/s
and $dP = dh\rho g = 25.8 \times 1000 \times 9.81 = 253\,098$ Pa
thus power $= 0.0994 \times 253\,098 = 25\,158$ W
and power $= 25.2$ kW.

Solution (b)
The Bernoulli equation is easily adapted to frictional flow and applying it here to sections A and B, Figure 7.5,

$$Z_a + (P_a/\rho g) + (u_a^2/2g) = Z_b + (P_b/\rho g) + (u_b^2/2g) + \text{loss}$$

substituting $30 + 0 + 0 = 0 + 0 + (u_b^2/2g) + \text{loss}$
from which frictional loss $= 30 - 25.8 = 4.2$ m of water.

Solution (c)
The efficiency of power transmission = energy at the nozzle/energy at the reservoir. The energy at the nozzle is in the form of kinetic energy and the energy in the water stored in the high level reservoir is in the form of potential energy.

Thus efficiency $= 25.8/30 = 0.86 = 86\%$

Summary for Example 7.5
With a pressure equivalent to 2.5 bar, you will notice the significant value of the power transmission. It confirms the reason why more

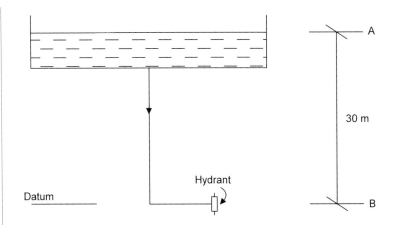

Figure 7.5 Example 7.5.

than one fireman may be required to hold the nozzle steady when it is connected by canvas hose to the hydrant. The losses due to friction will account for the loss at the exit from the reservoir, the loss in straight pipe and the loss in pipe fittings. It will be shown below that there is no shock loss at the nozzle.

Frictional losses in pipes, ducts and fittings

Frictional losses in pipe and duct systems may therefore include the following:

- shock losses
- losses in the straight pipe or duct
- losses in fittings including bends, tees, valves, volume control dampers etc.
- manufacturers of items of plant will provide the loss due to friction at given flow rates.

Shock loss usually occurs at sudden enlargements and sudden contractions. The entry to and exit from a large vessel such as the flow and return connections on the secondary side of a hot water service calorifier provides one example.

In the case of air flow, shock loss occurs across a supply air diffuser and a return air grille. For a sudden enlargement frictional loss $dh = (u_1 - u_2)^2/2g$ m of fluid flowing. A special case occurs when water discharges into a large tank of water or air is discharged into a room. In these cases u_2 approaches zero and therefore $dh = u_1^2/2g$.

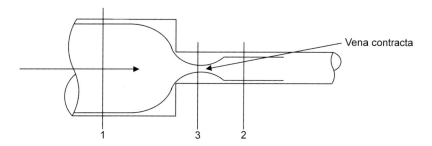

Figure 7.6 The sudden contraction.

A further special case occurs when water discharges into air. In this example $u_1 = u_2$ and therefore dh = zero.

- For a sudden contraction, frictional loss $dh = 0.5(u_2^2/2g)$ m of fluid flowing.
- For frictional losses in straight pipes and ducts in which the fluid flow is turbulent the Darcy equation applies and $dh = 4fLu^2/2gd$ m of fluid flowing.
- Frictional losses in fittings $dh = k(u^2/2g)$ m of fluid flowing. k = the velocity head loss factor for the fitting. Since it is dimensionless it has the same value as the velocity pressure loss factor.

Typical values of k are given for a variety of pipe and duct fittings in Section C4 of the *CIBSE Guide* where it is identified as Greek letter ζ.

The head loss for a sudden contraction given above is dependent upon the coefficient of contraction C_c at the fluid's vena contracta downstream of the fitting (Figure 7.6);
$C_c = a_3/a_2$ and $dh = u_2^2/2g\,[(1/C_c) - 1]^2$. The term $[(1/C_c) - 1]^2$ reduces to 0.5 when $C_c = 0.585$.

It is usually assumed that $C_c = 0.585$ and hence $dh = 0.5(u_2^2/2g)$ is taken for most sudden contractions, due to the difficulty in determining the coefficient of contraction by experiment for each fitting. u_1 refers to the mean velocity of the fluid upstream of the fitting and u_2 refers to the mean velocity of the fluid downstream of the fitting.

A final special case relates to the use of the bellmouth at the entry into or exit from a large vessel in the case of water or exit from/entry into a room in the case of air flow. The **bellmouth** replaces the sharp edge of the entry or exit point with a radiused 'edge'. This has the effect of reducing the shock loss to zero.

There now follow some examples to which apply the rational formulae introduced in this section.

Example 7.6
A horizontal pipe transporting water at 12 l/s suddenly increases from 100 to 200 mm in bore. Determine the shock loss in metres of water flowing, in units of pressure and units of energy.

Take the density of water as 1000 kg/m^3.

Solution
For a sudden enlargement $dh = (u_1 - u_2)^2/2g$ m of water flowing,

$u_1 = Q/a = 0.012 \times 4/\pi \times (0.1)^2 = 1.528$ m/s

$u_2 = 0.012 \times 4/\pi \times (0.2)^2 = 0.382$ m/s

substituting: $dh = (1.528 - 0.382)^2/2g = 0.067$ m of water flowing

in units of pressure: $dP = dh\rho g = 0.067 \times 1000 \times 9.81 = 657$ Pa
in units of energy: $dE = dhmg = 0.067 \times 12 \times 9.81 = 7.89$ Nm (J)

Example 7.7
A horizontal pipe carrying oil suddenly increases from 80 mm to 150 mm in bore. The fluid displacement in a differential manometer connected either side of the enlargement is 18 mm.

(a) Determine the shock loss and flow rate of oil.
Take the density of oil as 935 kg/m^3 and that for the measuring fluid as 13 600 kg/m^3.
(b) What is the velocity head loss factor for the fitting?

Solution (a)
From Chapter 5 the equivalent displacement of oil $dh = x(S - 1) = 0.018[(13\,600/935) - 1]$ from which $dh = 0.244$ m of oil flowing $= (P_2 - P_1)/\rho g$. Note this is not the shock loss for the enlargement; it is the regain in static head which results from a decrease in velocity.

Section 1 is upstream and section 2 is downstream of the sudden enlargement, for steady flow $Q_1 = Q_2$ and therefore $u_1 a_1 = u_2 a_2$ from which $u_2 = u_1 \times a_1/a_2$, thus $u_2 = u_1 \times (d_1/d_2)^2$; substituting $u_2 = u_1 \times (80/150)^2 = 0.285 u_1$.

Adopting the Bernoulli equation for frictional flow in which $Z_1 = Z_2$ and the frictional loss for a sudden enlargement $dh = (u_1 - u_2)^2/2g$
thus

$$(P_1/\rho g) + (u_1^2/2g) = (P_2/\rho g) + (u_2^2/2g) + (u_1 - u_2)^2/2g$$

rearranging

$$(P_2 - P_1)/\rho g = (u_1^2 - u_2^2)2g - (u_1 - u_2)^2/2g.$$

Remember that for an enlargement there is a gain in static head, known as static regain, and a loss in velocity head, thus $P_2 > P_1$ and $u_1 > u_2$.

Expanding the right-hand side
$$(P_2 - P_1)/\rho g = [(u_1^2 - u_2^2) - (u_1^2 - 2u_1u_2 + u_2^2)]/2g$$
thus $(P_2 - P_1)/\rho g = (2u_1u_2 - 2u_2^2)/2g$
$$= 2u_2(u_1 - u_2)/2g = \text{regain in static head}$$
substituting:
$$0.244 = [2 \times 0.285u_1(u_1 - 0.285u_1)]/2g$$
$$0.244 \times 2g = (0.40755\,u_1^2)$$
$$u_1^2 = (0.244 \times 2/0.040755)^{0.5} = 11.974$$
$$u_1 = 3.46\,\text{m/s}$$

shock loss: $dh = (3.46 - 0.285 \times 3.46)^2/2g = 0.312\,\text{m of oil flowing}$

flow rate: $Q = 3.46 \times \pi \times (0.08)^2/4 = 0.0174\,\text{m}^3/\text{s} = 17.4\,\text{l/s}$.

Solution (b)
The velocity head loss factor is normally based upon the larger of the two fluid velocities, thus from $dh = k(u^2/2g)$ the velocity adopted will be 3.46 m/s and
$$0.312 = k(3.46^2)/2g$$
from which $k = 0.511$.

Example 7.8
A horizontal pipe carrying water suddenly contracts from 300 to 100 mm bore. A differential manometer is connected either side of the sudden contraction and the pressure drop is 8.34 kPa.

(a) Determine the shock loss and the flow rate.
    Take the density of water as 1000 kg/m³.
(b) What is the velocity head loss factor for the fitting?

Solution (a)
Adopting the Bernoulli equation for frictional flow and given $Z_1 = Z_2$:
$$(P_1/\rho g) + (u_1^2/2g) = (P_2/\rho g) + (u_2^2/2g) + 0.5u_2^2/2g$$
rearranging and remembering that $P_1 > P_2$ and $u_2 > u_1$:
$$(P_1 - P_2)/\rho g = [(u_2^2 - u_1^2)/2g] + 0.5u_2^2/2g.$$
Now $u_2 = u_1(d_1/d_2)^2 = u_1(300/100)^2 = 9u_1$:
substituting:
$$8340/1000g = [(9u_1)^2 - u_1^2]/2g + 0.5 \times (9u_1)^2/2g$$

$0.85 \times 2g = 80u_1^2 + 40.5u_1^2$

$16.68 = 120.5u_1^2$

from which $u_1 = 0.372$ m/s

shock loss $\quad dh = 0.5(9 \times 0.372)^2/2g = 0.286$ m of water flowing

flow rate $\quad Q = 0.372 \times \pi \times (0.3)^2/4 = 0.0263$ m^3/s

thus $\quad Q = 26.3$ l/s.

Solution (b)
The velocity head loss factor is normally based upon the larger of the two fluid velocities and is obtained from the equation $dh = k(u^2/2g)$ m of fluid flowing.

Substituting the shock loss: $0.286 = k(3.46^2/2g)$ and therefore the velocity head loss factor k for the sudden contraction is 0.469.

A special case

When water flows from a supply pipe into a large tank or when air flows from a supply duct into a room the shock loss $dh = u_1^2/2g$ m of fluid flowing where u is the fluid velocity in the pipe or duct. Since the frictional loss is also expressed as $dh = k(u^2/2g)$ then $(u_1^2/2g) = k(u^2/2g)$ therefore the velocity head loss factor k must equal 1.0. Remember that frictional losses in pipe and duct fittings are expressed as fractions of the velocity head: $(u^2/2g)$ or velocity pressure: $(\rho u^2/2)$, thus:

$$dP = k(\rho u^2/2)\text{ Pa} \quad \text{and} \quad dh = k(u^2/2g)\text{ m of fluid flowing.}$$

Since the term k is dimensionless it can be used in either of these formulae. It can be defined as the velocity head loss factor or the velocity pressure loss factor.

Example 7.9

The pressure loss across a globe valve located in a horizontal pipe is measured as 126 mbar when flow velocity is 1.9 m/s. Determine the velocity pressure loss factor given water density as 1000 kg/m^3.

Solution

Now $dP = k\rho u^2/2$.

Adopting the Bernoulli equation for frictional flow taking sections 1 and 2 as upstream and downstream of the valve respectively:

$$Z_1 + (P_1/\rho g) + (u_1^2/2g) = Z_2 + (P_2/\rho g) + (u_2^2/2g) + k(u_2^2/2g).$$

Since the valve is horizontal $Z_1 = Z_2$, water velocity either side of the valve is the same, thus $u_1 = u_2$.

Thus the Bernoulli equation reduces to: $(P_1 - P_2)/\rho g = k(u^2/2g)$ m of water. Note therefore that if there is no change in fluid velocity either side of the fitting the frictional loss through it is equal to the static pressure drop. This is not the case for a sudden enlargement or a sudden reduction where a change in velocity does occur either side of the fitting. See Examples 7.7 and 7.8 in which the fluid velocities u_1 and u_2 in the Bernoulli equation are not the same and therefore do not cancel.

Thus for the globe valve $(P_1 - P_2)/\rho g = dh = k(u^2/2g)$ m of water or $dP = k(\rho u^2/2)$ Pa.

Substituting: $12\,600 = k \times 1000 \times 1.9^2/2$

from which the velocity pressure loss factor for the globe valve $k = 6.98$.

Example 7.10
The displacement of measuring fluid in a differential manometer connected either side of a 50 mm bore gate valve located horizontally is 2 mm. If the mass flow rate of water at 75°C through the valve is 2.87 kg/s determine its velocity head loss factor. The density of measuring fluid is 13 600 kg/m³.

Solution
At 75°C water density is 975 kg/m³. (*Refer to the Thermodynamic and Transport Properties of Fluids.*)

The corresponding displacement of water in the manometer can be determined from $dh = x(S - 1)$ m of water flowing (Chapter 5), thus $dh = 0.002[(13\,600/975) - 1] = 0.0259$ m of water flowing.

The volume flow rate of water flowing $Q = 2.87/975 = 0.00294$ m³/s.

Since $Q = u \times a$, mean velocity $u = Q/a = 0.00294 \times 4/\pi(0.05)^2 = 1.5$ m/s. Since the fluid velocity either side of the valve is the same, $u_1 = u_2$, and as the pipe is horizontal $Z_1 = Z_2$, the Bernoulli equation is reduced to:

$(P_1 - P_2)/\rho g = dh = k(u^2/2g)$

and substituting

$0.0259 = k \times 1.5^2/2g$

from which the velocity head loss factor for the gate valve $k = 0.226$.

Example 7.11
A section of a heating system in which water flows at 0.35 kg/s comprises:

1 column radiator $k = 5$
2×20 mm angle radiator valves $k = 5$ each
7×20 mm malleable cast iron bends $k = 0.7$ each
8 m × 20 mm bore black mild steel pipe in which the coefficient of friction $f = 0.0045$. Determine the pressure loss due to friction given water density as 975 kg/m³.

Solution
Volume flow of water $Q = 0.35/975 = 0.000\,359\,\text{m}^3/\text{s}$.

Mean water velocity $u = Q/a = 0.000\,359 \times 4/\pi(0.02)^2$
$= 1.143\,\text{m/s}$.

The total velocity pressure loss factor for the fittings in the pipe section $k_t = 19.9$. Assuming no change in fluid velocity in the pipeline and ignoring changes in height, the Bernoulli equation for the fittings is reduced to: $dP = k(\rho u^2/2)$ Pa. Then by substitution: the pressure loss attributable to the fittings

$dP = 19.9 \times 975 \times 1.143^2/2 = 12\,674\,\text{Pa}$.

Similarly for head loss in straight pipe assuming turbulent flow the Bernoulli equation is reduced to: $dh = 4fLu^2/2gd$ m of water,

substituting: $dh = 4 \times 0.0045 \times 8 \times 1.143^2/2g \times 0.02 = 0.48$ m of water flowing

then $dP = dh\rho g = 0.48 \times 975 \times 9.81$

from which the pressure loss in the straight pipe $dP = 4591$ Pa.

The total hydraulic pressure loss in the pipe section $dP = 12\,674 + 4591 = 17\,265$ Pa.

Example 7.12
Water flows at 30 kg/s in a 100 m bore straight pipe 68 m in length. Determine:

(a) The head loss due to friction given water density as 1000 kg/m³, the coefficient of friction f as 0.004 and viscosity μ as 0.000 015 01 kg/ms.
Confirmation should be sought that flow is in the turbulent region.
(b) The gradient to which the pipe must be laid to maintain a constant head.

Solution (a)

Volume flow $Q = M/\rho = 30/1000 = 0.03 \text{ m}^3/\text{s}$.

Mean velocity $u = Q/a = 0.03 \times 4/\pi \times 0.1^2 = 3.82 \text{ m/s}$.

Reynolds number $Re = \rho u d/\mu = 1000 \times 3.82 \times 0.1/0.000\,015\,01$
$= 25\,466\,667$.

Since $Re > 3500$ flow is confirmed as being in the turbulent region. The Darcy equation for turbulent flow in straight pipes $dh = 4fLu^2/2gd$, and substituting $dh = 4 \times 0.004 \times 68 \times 3.82^2/2 \times 9.81 \times 0.1 = 8.09 \text{ m}$ of water. Head loss due to friction $dh = 8.09 \text{ m}$ of water over a straight pipe length of 68 m.

Solution (b)
Refer to Figure 7.7.

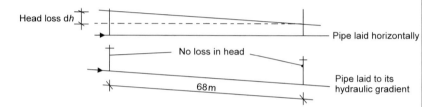

Figure 7.7 The hydraulic gradient (Example 7.12).

If the pipe is laid to a gradient such that it terminates at a point 8.09 m vertically below the point from which it started the gradient of $8.09/68 = 0.119 \text{ m/m}$ will ensure that the head loss due to friction is offset by the gradient which is $1/0.119 = 1$ in 8.4. This is called the hydraulic gradient of the pipe.

Example 7.13

Determine the regain in static pressure for the transformation piece shown in Figure 7.8 if the velocity pressure loss factor for the fitting is 0.25.

The volume flow rate of air at a temperature of 28°C is $3 \text{ m}^3/\text{s}$ and the duct increases gradually in diameter from 0.6 m to 1.0 m.

Take air density as 1.2 kg/m^3 at 20°C.

Solution
For steady flow $Q = u_1.a_1 = u_2.a_2$.

$u_1 = Q/a_1 = 3 \times 4/\pi \times 0.6^2 = 10.61 \text{ m/s}$
$u_2 = Q/a_2 = 3 \times 4/\pi \times 1^2 = 3.82 \text{ m/s}$.

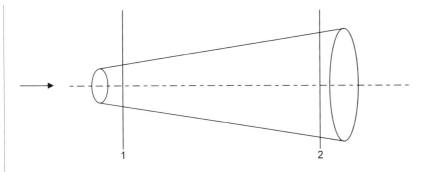

Figure 7.8 Example 7.13.

Adopting the Bernoulli equation for frictional flow at sections 1 and 2

$(P_1/\rho g) + (u_1^2/2g) = (P_2/\rho g) + (u_2^2/2g) + k(u_1^2/2g)$.

Remembering that for an enlargement $P_2 > P_1$ and $u_1 > u_2$,
rearranging: $(P_2 - P_1)/\rho g = (u_1^2 - u_2^2)/2g - k(u_1^2/2g)$

substituting:

$$(P_2 - P_1)/\rho g = [(10.61^2 - 3.82^2)/2g] - (0.25 \times 10.61^2/2g)$$
$$= 4.994 - 1.434$$
$$= 3.56 \text{ m of air flowing.}$$

Absolute air temperature in the duct $= 273 + 28 = 301$ K.

Absolute air temperature at 1.2 kg/m^3 and $20°$C $= 273 + 20 = 293$ K.

Air density correction $\rho = 1.2(293/301) = 1.168$ kg/m^3 at $28°$C

$$dP = 3.56 \times 1.168 \times 9.81 = 41 \text{ Pa.}$$

The static pressure regain generated by the transformation piece is 41 Pa. This is distributed along the duct downstream of the fitting.

Example 7.14
A ground storage tank supplies water to a high-level tank in a building the vertical distance being 25 m. A multistage centrifugal pump is installed on the suction pipe from the low-level tank and discharges 2.5 kg/s of water into the 50 mm bore rising main which terminates with a ball valve at the high-level tank.

Ignore the effects of pressure in the pump suction and determine the net pump duty and output power.

Data: Velocity pressure loss factors: 2 bends $k = 0.4$ each, 2 stop valves $k = 0.7$ each, 1 recoil valve $k = 8.0$.

146 Flow of fluids in pipes, ducts and channels

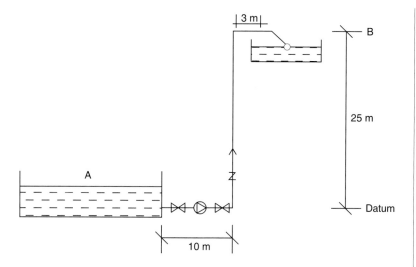

Figure 7.9 Example 7.14.

Assuming turbulent flow, the coefficient of friction in the straight pipe $f = 0.005$. Pressure required at the ball valve = 30 kPa. Density of water is 1000 kg/m³.

Solution
The system is shown in Figure 7.9.
Adopting the Bernoulli equation for frictional flow and considering sections a and b

$$Z_a + (P_a/\rho g) + (u_a^2/2g) = Z_b + (P_b/\rho g) + (u_b^2/2g) + \text{losses}$$

now $P_a = P_b$ = atmospheric pressure and u_a approaches zero velocity.

Placing all the terms remaining from the Bernoulli equation on to the right-hand side will represent the total energy required to move 2.5 kg/s of water from point a to point b. The net pump head dh that must therefore be generated will be:

$$dh = (Z_b - Z_a) + (u_b^2/2g) + \text{losses}.$$

The losses include those through the fittings, that through the straight pipe and the discharge pressure required at the ball valve which can be added at the completion of the solution.

Thus $dh = (Z_b - Z_a) + (u_b^2/2g) + k(u_b^2/2g) + 4fLu_b^2/2gd$.

The total velocity head loss factor is 10.2. You should now confirm that this is so.

Mean velocity of flow $u_b = Q/a = 0.0025 \times 4/\pi \times 0.05^2$
$= 1.273$ ms.

Substituting:

$$dh = (25 - 0) + (1.273^2/2g) + 10.2(1.273^2/2g) + [(4 \times 0.005 \\ \times 38 \times 1.273^2)/(2g \times 0.05)]$$

from which $dh = 25 + 0.0826 + 0.8425 + 1.2556$

and $dh = 27.18$ m of water flowing

$dP = 27.18 \times 1000 \times 9.81 = 266\,643$ Pa.

Net pump pressure $= 267 + 30 = 297$ kPa
and the net pump duty is 2.5 kg/s at 297 kPa.
The pump output power $= dhMg = (297\,000/1000 \times 9.81) \times 2.5 \times 9.81 = 743$ W.

Example 7.15
Two high-level cold water storage tanks having a capacity of 4500 l each are refilled every two hours. The vertical height of the supply water main will be 26 m and its horizontal distance from the water utilities main is 9 m.

If the available pressure is 300 kPa during peak demand size the rising main.

Data: assuming turbulent flow, the coefficient of friction $f = 0.007$, allowance for fittings is 30% on the straight pipe, pressure required at the index ball valve is 30 kPa.

Solution
Figure 7.10 shows the system in elevation.
The mass flow rate $M = (4500 \times 2)/(2 \times 3600) = 1.25$ kg/s.
Adopting the Bernoulli equation for frictional flow

$$Z_a + (P_a/\rho g) + (u_a^2/2g) = Z_b + (P_b/\rho g) + (u_b^2/2g) + \text{loss}.$$

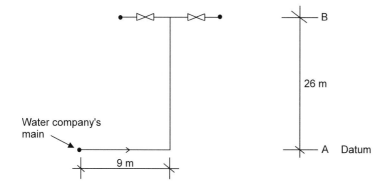

Figure 7.10 Example 7.15.

Assuming $u_a = u_b$ and rearranging the equation knowing that $P_a > P_b$

$$(Z_a - Z_b) + (P_a - P_b)/\rho g = \text{loss}$$

From Chapter 6 Box's formula for head loss in turbulent flow in straight pipe can be adopted here, thus

$$(P_a - P_b)/\rho g + Z_a - Z_b = fLQ^2/3d^5.$$

Total equivalent length of pipe including fittings $L = (26 + 9) \times 1.3 = 45.5$ m.

Substituting:

$$[(300\,000 - 30\,000)/1000 \times 9.81] + 0 - 26 = [0.007 \times 45.5 \\ \times (0.001\,25)^2]/(3 \times d^5)$$

thus $27.52 - 26 = 1.6589 \times 10^{-7}/d^5$

from which $d^5 = 1.52 \times 1.6589 \times 10^{-7} = 2.5215 \times 10^{-7}$

and $d = 0.0479$ m $= 47.9$ mm.

The nearest standard pipe diameter for the rising main $= 50$ mm.

7.4 Semi-graphical solutions to frictional flow in pipes and ducts

Solutions to problems involving the flow of water in pipes can be undertaken using the Moody chart of Poiseuille and Colebrook–White. This chart is reproduced here from the *CIBSE Guide* to current practice by kind permission of CIBSE (Figure 6.13). The calculation routine is as follows: determine the Reynolds number for the known flow conditions from $Re = \rho u d/\mu$, determine the pipe roughness ratio k_s/d, determine the value of the frictional coefficient f from the Moody chart, if the flow is turbulent determine the head loss due to friction from the Darcy equation $dh = 4fLu^2/2gd$. Table 7.1 lists values of absolute roughness k_s, which is a lineal measurement of high points on the rough internal surface, for pipes and ducts.

Table 7.1 Values of absolute roughness in pipes and ducts. (Reproduced from *CIBSE Guide* section C4 (1986) by permission of the Chartered Institution of Building Services Engineers.)

Pipes and ducts	k_s (mm)
Copper pipe	0.0015
Plastic pipe	0.003
New black steel pipe	0.046
Rusted black steel pipe	2.5
Clean aluminium ducting	0.05
Clean galvanized ducting	0.15
New galvanized pipe	0.15

Example 7.16

A horizontal straight pipe 32 mm bore by 25 m in length carries water at the rate of 1 kg/s. If it is new black steel pipe determine the head loss in metres of water, the pressure loss in Pascals and the specific pressure loss in Pa/m.

Data: absolute viscosity 0.000 378 N s/m^2, density 975 kg/m^3.

Solution

In order to use the Moody chart to obtain the coefficient of friction the relative roughness or roughness ratio and the Reynolds number must be found.

Roughness ratio $= k_s/d = 0.046/32 = 0.001\,44$.

Note that pipe diameter d is left in mm since k_s is measured in mm. From Chapter 6 the Reynolds number

$$Re = dM/\mu a = [(0.032 \times 1 \times 4)/(0.000\,378 \times \pi \times 0.032^2)]$$
$$= 105\,261$$

given Re and k_s/d, the coefficient of friction f can now be found from the Moody chart, Figure 6.13, from which $f = 0.0055$.

As the flow exceeds the Reynolds number of 3500 flow is clearly turbulent and Darcy's equation can be used.

Thus $dh = 4fLu^2/2gd$ m of water flowing

where $u = M/\rho a = (1 \times 4)/(975 \times \pi \times 0.032^2) = 1.275$ m/s

and therefore $dh = [(4 \times 0.0055 \times 25 \times 1.275^2)/(2 \times 9.81 \times 0.032)]$

$dh = 1.424$ m of water.

Pressure loss $dP = dh\rho g = 1.424 \times 975 \times 9.81 = 136\,215$ Pa.

Specific pressure loss $dp = dP/L = 13\,621/25 = 544$ Pa/m.

Example 7.17

Air at a temperature of 8°C flows at 1.02 m^3/s along a straight galvanized duct 400 mm in diameter. Determine the specific static pressure loss in Pa/m.

Take air density at 20°C as 1.2 kg/m^3 and air viscosity at 8°C as 0.000 017 55 kg/ms.

Solution

Reynolds number $Re = \rho u d/\mu$.

The viscosity of dry air which is given here can be obtained from the tables of *Thermodynamic and Transport Properties of Fluids*.

Corrected air density $\rho = 1.2[(273 + 20)/(273 + 8)] = 1.251 \text{ kg/m}^3$
mean air velocity $u = Q/a = 1.02 \times 4/\pi \times 0.4^2 = 8.12 \text{ m/s}$
from which $Re = (1.251 \times 8.12 \times 0.4)/(0.00001755) = 232\,186$.
Relative roughness $= k_s/d = 0.15/400 = 0.000\,375$.
From the Moody chart, Figure 6.13, the coefficient of friction $f = 0.0044$.

Since flow is fully turbulent, Darcy's equation can be used and

$dh = 4fLu^2/2gd$ where $L = 1.0 \text{ m}$.

Substituting: $dh = (4 \times 0.0044 \times 1 \times 8.12^2)/(2 \times 9.81 \times 0.4)$

from which for one metre of straight duct $dh = 0.1479 \text{ m}$ of air flowing.

and specific pressure loss $dp = h\rho g/L = (0.1479 \times 1.251 \times 9.81)/1 = 1.815 \text{ Pa/m}$.

Summary for Example 7.17
An assumption is made in this solution that the air flowing in the duct is not subject to compression. This is acceptable in ventilation and air conditioning systems but the assumption cannot be made when working with compressed air because of the influence of pressure on air density.

7.5 Gravitational flow in flooded pipes

Water flow subject to gravity will occur, for example, from a tank located at high level which supplies water to a point at a lower level. It also occurs when a high-level reservoir supplies water to a reservoir at some lower level. Without a prime mover such as a pump, the gradient or vertical drop through which the pipe is routed offsets exactly the force of friction opposing fluid flow.

Example 7.18
A pipe with a constant gradient connects two reservoirs having a difference in water levels of 20 m. The upper 200 m of pipe is 100 mm bore, the next 100 m is 200 mm bore and the final 100 m is 100 mm bore. The pipes all have coefficients of friction of 0.006. The connections to the reservoirs are bellmouthed and the changes in pipe bore are sudden. Determine the flow of water through the connecting pipe.

Solution
Figure 7.11 shows the system in elevation.
Adopting the Bernoulli equation for frictional flow

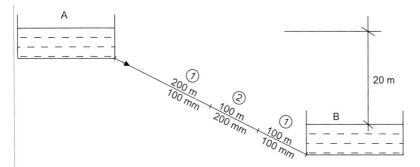

Figure 7.11 Example 7.18.

$$Z_a + (P_a/\rho g) + (u_a^2/2g) = Z_b + (P_b/\rho g) + (u_b^2/2g) + \text{losses}$$

$P_a = P_b = $ atmospheric pressure, $u_a = u_b$

thus $Z_a - Z_b = $ losses = loss in pipe 1 + enlargement loss
 + loss in pipe 2 + reduction loss + loss in pipe 1.

Since the exit and entry to the reservoirs are bellmouthed the shock losses due to friction are negligible.

Thus $20 = (4fL_1u_1^2/2gd_1) + [(u_1 - u_2)^2/2g] + (4fL_2u_2^2/2gd_2)$
 $+ (0.5u_1^2/2g) + (4fL_1u_1^2/2gd_1)$

for steady flow $Q = u_1a_1 = u_2a_2$

and $u_1 = (a_2/a_1)u_2 = (d_2/d_1)^2 u_2 = (200/100)^2 u_2$

from which $u_1 = 4u_2$

Substituting for u_1 and also the other data we have:

$20 = 39.1u_2^2 + 0.46u_2^2 + 0.61u_2^2 + 0.41u_2^2 + 19.6u_2^2$

$20 = u_2^2(39.1 + 0.46 + 0.61 + 0.41 + 19.6)$

$20 = 60.18u_2^2$

from which $u_2 = 0.576$ m/s.

$Q = ua = 0.576 \times \pi \times 0.2^2/4 = 0.0181$ m³/s

thus the gravitational flow rate = 18.1 l/s.

Gravitational mass transfer of water = 18.1 kg/s.

Example 7.19

Figure 7.12 shows a pipe system connecting two tanks in which the entry and exit losses may be ignored since the connections are bellmouthed. The velocity head loss factor for the bends is 0.3.

152 Flow of fluids in pipes, ducts and channels

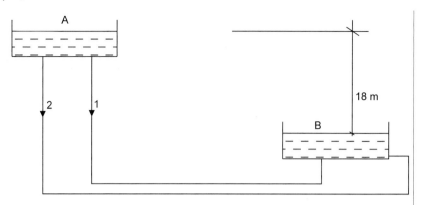

Figure 7.12 Example 7.19.

Determine the total flow rate from the tank at level A to the tank at B.
Data: pipe 1 is 70 m long and 65 mm bore, f is 0.005
pipe 2 is 80 m long and 50 mm bore, f is 0.006.

Solution
Adopting the Bernoulli equation for frictional flow, $P_a = P_b =$ atmospheric pressure and $u_a = u_b$.
Thus $Z_a - Z_b =$ losses in pipe 1 $=$ losses in pipe 2 since the pipes are in parallel. Taking the losses in pipe 1

$$Z_a - Z_b = (4fL_1u_1^2/2gd_1) + k(u_1^2/2g)$$

rearranging: $2g(Z_a - Z_b) = u_1^2(4fL_1/d_1 + k)$
rearranging in terms of u_1:

$$u_1 = [2g(Z_a - Z_b)/((4fL_1/d_1) + 2k)]^{0.5}$$

substituting: $u_1 = [2g \times 18/((4 \times 0.005 \times 70/0.065) + 0.6)]^{0.5}$
$= 3.99 \, \text{m/s}$

also $u_2 = [2g \times 18/((4 \times 0.006 \times 80/0.05) + 0.9)]^{0.5} = 3.0 \, \text{m/s}$
For parallel flow $Q =$ flow in pipe 1 plus flow in pipe 2

$$Q_1 = u_1a_1 = 3.99 \times \pi \times 0.065^2/4 = 0.013\,24 \, \text{m}^3/\text{s}$$

and

$$Q_2 = u_2a_2 = 3.0 \times \pi \times 0.05^2/4 = 0.005\,89 \, \text{m}^3/\text{s}$$

therefore total gravitational flow $Q = Q_1 + Q_2 = 0.019\,13 \, \text{m}^3/\text{s}$
$= 19.13 \, \text{l/s}.$

Gravitational mass transfer of water $= 19.13 \, \text{kg/s}.$

Example 7.20
Two tanks, one 15 m vertically above the other as shown in Figure 7.13, are connected together by a 40 mm pipe. Determine the rate of flow of water from the upper to the lower tank.

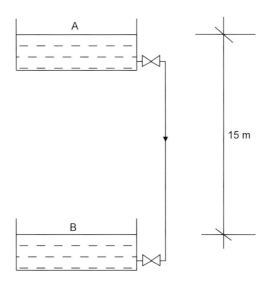

Figure 7.13 Example 7.20.

How long will it take to fill the lower tank, given its dimensions are $5 \times 2 \times 1$ m to the water level?

Data: head loss factors are, for bends 0.3, for stop valves 5.0.

Solution
Adopting the Bernoulli equation for frictional flow, $P_a = P_b$ = atmospheric pressure, u_a and u_b are approaching zero.

Rearranging the equation therefore $Z_a - Z_b$ = losses

thus $Z_a - Z_b$ = loss in sudden contraction + loss in pipe
+ loss in fittings + loss in sudden enlargement

and $Z_a - Z_b = (0.5u^2/2g) + (4fLu^2/2gd) + (ku^2/2g) + u^2/2g$.

Note that for the sudden enlargement the water velocity in the tank approaches zero thus the shock loss reduces to $u^2/2g$.

Thus simplifying: $Z_a - Z_b = (u^2/2g)[0.5 + (4fL/d) + \Sigma k + 1]$
substituting: $15 = (u^2/2g)[0.5 + (4 \times 0.005 \times 15/0.04) + 10.6 + 1]$
from which $u = 3.875$ m/s.
Flow rate will be $Q = ua = 3.875 \times \pi \times 0.04^2/4 = 0.00487 \, \text{m}^3/\text{s}$

gravitational flow $Q = 4.87$ l/s
gravitational mass transfer of water $= 4.87$ kg/s
time to fill the tank $=$ volume of tank to the waterline/flow rate
$$= 5 \times 2 \times 1/0.00487 = 2053 \text{ s} = 34 \text{ min}.$$

Example 7.21
Figure 7.14 shows a pipe arrangement connected to a high-level tank by which water is discharged to atmosphere at points C and D. Given that the coefficient of friction is 0.005 for all the pipes and that the velocity head loss factor for each bend is 0.7 and that for each valve is 3.0, determine the flow rate at points C and D.
   Data: A–B = 15 m of 25 mm bore pipe, B–C = 21 m of 20 mm bore pipe, B–D = 15 m of 15 mm bore pipe.

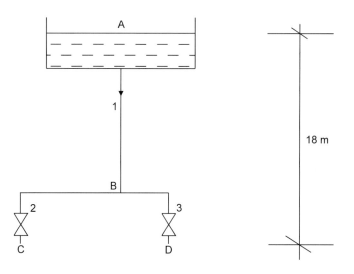

Figure 7.14   Example 7.21.

Solution
From Figure 7.14 pipe sections B–C and B–D are in parallel
thus $dh_{b-c} = dh_{b-d}$ and $dh_{1,2} = dh_{1,3}$
thus $(4fL_2 u_2^2/2gd_2) + k(u_2^2/2g) = (4fL_3 u_3^2/2gd_3) + k(u_3^2/2g)$
simplifying by cancelling $1/2g$
$$(4fL_2 u_2^2/d_2) + ku_2^2 = (4fL_3 u_3^2/d_3) + ku_3^2$$
$$u_2^2[(4fL_2/d_2) + k] = u_3^2[(4fL_3/d_3) + k]$$

substituting:

$$u_2^2[(4 \times 0.005 \times 21/0.02) + 3.7] = u_3^2[(4 \times 0.005 \times 15/.015) + 3.7]$$

from which $\qquad 24.7u_2^2 = 23.7u_3^2$

thus $\qquad u_2^2 = (23.7/24.7)u_3^2$

and therefore $\qquad u_2 = 0.9795u_3.$

For continuity of flow $Q_1 = Q_2 + Q_3$

then $\qquad u_1 \pi d_1^2/4 = (u_2 \pi d_2^2/4) + (u_3 \pi d_3^2/4)$

from which $\qquad u_1 d_1^2 = u_2 d_2^2 + u_3 d_3^2$

and $\qquad u_1 = (1/d_1^2)(u_2 d_2^2 + u_3 d_3^2)$

substituting for u_2:

$$u_1 = (1/d_1^2)(0.9795 u_3 d_2^2 + u_3 d_3^2)$$

substituting for d_1, d_2 and d_3:

$$u_1 = 1600(0.000\,392 u_3 + 0.000\,225 u_3)$$

therefore $u_1 = 0.9872 u_3$.

Now $dh_{a-c} = dh_{a-d}$ and $dh_{1,2} = dh_{1,3}$ since pipes 2 and 3 are in parallel. Working with pipe circuit consisting of sections 1 and 2 and adopting the Bernoulli equation for frictional flow in which $P_a = P_c =$ atmospheric pressure and u_a approaches zero. Note also that the head loss due to water discharge into air at point C is zero. Thus $Z_a - Z_c = (u_c^2/2g) +$ [losses]

The losses include: [sudden contraction + loss in pipe 1
+ sudden contraction at tee + loss in pipe 2
+ loss in bend and valve].

$$Z_a - Z_c = (u_2^2/2g) + [(0.5u_1^2/2g) + (4fL_1 u_1^2/2gd_1) + (0.5u_2^2/2g) + (4fL_2 u_2^2/2gd_2) + k(u_2^2/2g)].$$

substituting $u_2 = 0.9795u_3$, $u_1 = 0.9872u_3$, $k = 0.7 + 3 = 3.7$, $f = 0.005$, $L_1 = 15$ m, $L_2 = 21$ m, $d_1 = 0.025$ m, $d_2 = 0.02$ m:

$$18 = 1/2g[0.9795u_3^2 + 0.4873u_3^2 + 11.6948u_3^2 + 0.4797u_3^2 + 20.1478u_3^2 + 3.55u_3^2]$$

$$18 = (1/2g) \times 37.3391 u_3^2$$

from which

$u_3 = 3.075$ m/s

$u_1 = 0.9872 u_3 = 0.9872 \times 3.075 = 3.035$ m/s

$u_2 = 0.9795 u_3 = 0.9795 \times 3.075 = 3.012$ m/s

$$Q_c = u_2.a_2 = 3.012 \times \pi \times 0.02^2/4 = 0.0009462 \, \text{m}^3/\text{s}$$
$$= 0.95 \, \text{litres/s}$$
$$Q_d = u_3.a_3 = 3.075 \times \pi \times 0.015^2/4 = 0.0005434 \, \text{m}^3/\text{s}$$
$$= 0.54 \, \text{litres/s}$$

Gravitational mass transfer of water at points C and D are 0.95 kg/s and 0.54 kg/s respectively.

7.6 Gravitational flow in partially flooded pipes and channels

This section will consider the mass transfer of water in open channels and flooded and partially flooded pipes set to gradients to maintain flow.

When there is no prime mover such as a pump, fluid flow relies on the hydraulic gradient to which the pipe or channel is laid. The gradient offsets exactly the forces of friction opposing flow and generated by the moving fluid.

The Chezy formula is usually associated with this work and this is an adaption of the Darcy equation for turbulent flow in pipes where:

$dh = 4fLu^2/2gd$ metres of fluid flowing

The formula can be rearranged in terms of mean flow velocity u thus:

$u^2 = 2gddh/4fL$ m/s.

It can then be separated into constituent parts:

$u = (2g/f)^{0.5} \times (dh/L)^{0.5} \times (d/4)^{0.5}$ m/s.

The term in the Darcy equation which can vary in partial flow is $(d/4)$.
The constituent parts can then be identified thus:

$(2g/f)^{0.5} = C$, the Chezy coefficient.

When the coefficient of friction $f = 0.0064$, $C = 55$.
You should confirm this value for C.

$dh/L = i$ = the hydraulic gradient

$d/4 = m$ = the hydraulic mean diameter for full and half full bore flow in metres.

The hydraulic mean diameter $(d/4)$ will vary according to the volume of partial flow.

The determination of m is given below for various flow conditions.

m = (cross-sectional area of flow)/(length of wetted perimeter)

for full bore flow in conduits of circular cross-section
$m = (\pi d^2/4)/\pi d = d/4$

for half full bore flow in conduits of circular cross-section
$m = (\pi d^2/8)/(\pi d/2) = d/4$

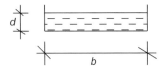

Figure 7.15 Hydraulic mean depth in rectangular channels $m = \dfrac{db}{2d+b}$.

for three-quarter full bore flow in conduits of circular cross-section it can be shown that $m = d/3$
for rectangular channels (Figure 7.15) hydraulic mean depth $m = db/(2d+b)$.

The term m in the Chezy formula must therefore be selected according to the cross-sectional shape of the conduit and the volume of the partial flow.

Thus the Chezy formula is given as $u = C(mi)^{0.5}$ m/s.
Invariably the Chezy coefficient is taken as $C = 55$.

Example 7.22
Determine the discharge capacity of a 160 mm drain flowing half full bore when it is laid to a gradient of 1:150. Take the Chezy coefficient C as 55.

Solution
The drain may be taking rainwater run off or it may be connected to a vertical soil stack. In either case the gradient is required to ensure flow to the point of discharge.
From the Chezy formula $u = 55(mi)^{0.5}$
substituting, mean flow velocity $u = 55[(0.16/4)(1/150)]^{0.5} = 55 \times 0.01633 = 0.898$ m/s. For half full bore flow in the drain

$Q = ua = 0.898 \times \pi(0.16)^2/8 = 0.009 \, \text{m}^3/\text{s}$
$Q = 9 \, \text{l/s}$.

Gravitational mass transfer of water $= 9$ kg/s.

Example 7.23
(a) Determine the gradient required for a 110 mm drain to run three-quarter full bore at a mean soil water velocity of 1.2 m/s. The coefficient of friction f for the drain pipe is 0.008.
(b) Determine the mass transfer of soil water.

Solution (a)

The Chezy coefficient $C = (2g/f)^{0.5} = (2g/0.008)^{0.5} = 49.523$.

From the Chezy formula $u^2 = C^2 \times m \times i$

rearranging: $i = u^2/C^2 m$ where for three-quarter full bore $m = d/3$

substituting: $i = 1.2^2 \times 3/49.523^2 \times 0.11 = 0.016\,013$ m/m

now the hydraulic gradient $i = dh/L = 0.016\,013$

therefore for a fall of one metre, drain length $L = dh/0.016\,013 = 1/0.016\,013 = 62.4$.

Thus the minimum hydraulic gradient for the drain is 1:62.4

Alternative solution to part (a) of Example 7.23

The solution to this problem can be done by adopting the Bernoulli equation for frictional flow in which the head loss sustained can be obtained from the Darcy equation. Consider two points along the drain:

$$Z_a + (P_a/\rho g) + (u_a^2/2g) = Z_b + (P_b/\rho g) + (u_b^2/2g) + \text{loss}$$

now $P_a = P_b$ and $u_a = u_b$

thus rearranging: $Z_a - Z_b = \text{loss}$

The Darcy equation for turbulent flow $dh = 4fLu^2/2gd$ m of water flowing, but the hydraulic mean radius for three-quarter full bore $m = d/3$ therefore $dh = 3fLu^2/2gd$ and considering one metre length of drain pipe and substituting: $Z_a - Z_b = (3 \times 0.008 \times 1.0 \times 1.2^2)/(2g \times 0.11)$ and the vertical fall $Z_a - Z_b = 0.016\,013$ m.

For a one metre vertical fall the length of the drain pipe $L = 1/0.016\,013 = 62.4$ m.

Thus the minimum gradient to which the pipe must be laid to achieve a mean velocity of 1.2 m/s will be 1:62.4.

Solution (b)

Flow rate for three-quarter full bore $Q = ua = 1.2 \times 0.75(\pi \times 0.11^2/4) = 0.00\,855$ m³/s.

Gravitational mass transfer of soil water = 8.55 kg/s.

Flow in vertical soil stacks is subject to the full impact of gravitational acceleration. In such circumstances gravitational acceleration and opposing friction will balance and a terminal velocity is reached. For soil stacks which contain air and are open to atmosphere, in order to reduce water and air disturbance, stack loading is taken as

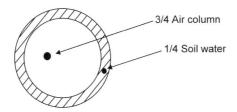

Figure 7.16 Vertical soil stack – accommodation of air and water.

about one-quarter full. Due to the coanda effect the water slides down the inside of the stack wall leaving a core of air in the centre (Figure 7.16).

The empirical formula for sizing the vertical soil stack when flowing one-quarter full is $Q = Kd^{(8/3)}$ l/s.

The constant $K = 0.000\,032$ for quarter full flow and
$\qquad\quad\; = 0.000\,052$ for one-third full flow.

Soil stack diameter d is in millimetres.

Example 7.24
The simultaneous discharge of soil water into a vertical stack is estimated as 6.5 l/s. Determine the size of the stack to accommodate the discharge.

Solution
From the empirical formula for flow in vertical soil stacks

$$d = (Q/K)^{(3/8)} = (6.5/0.000\,032)^{(3/8)} = 97.8\,\text{mm}.$$

The nearest standard stack diameter $d = 100$ mm.

Rainwater run-off depends upon the surface on which it lands. The amount of water which can be expected from any given surface depends upon:

- area of surface upon which rain is falling
- surface type
- whether the surface is level or sloping
- the rainfall intensity
- the rate of evaporation, which is seasonal.

Rate of run-off $Q = APR/3600 \times 1000\,\text{m}^3/\text{s}$.

The impermeability P of the surface depends upon its type. Refer to Table 7.2.

Table 7.2 Impermeability of different surfaces

Type of surface	average P
Watertight	0.9
Asphalt	0.875
Closely jointed stone	0.825
Macadam roads	0.435
Lawns	0.15
Woods	0.105

Example 7.25
A car park having an asphalt surface measures 50 m × 30 m and is laid to ensure adequate water run-off into a drainage channel running continuously along the length of the parking area. Determine the gradient and diameter of each of the main drainpipes connected to each end of the drainage channel assuming the high point is halfway along its length.

Data: rainfall intensity is to be taken as 75 mm/h, mean water velocity in the drainpipes is 1.1 m/s and the drains are to run three-quarter full bore. The Chezy coefficient $C = 55$.

Solution
Rainfall run-off $Q = APR/3600 \times 1000 = (50 \times 30) \times 0.875$
$\times 75/3600 \times 1000 = 0.027\,34\,\text{m}^3/\text{s}$.

Each drain must handle $0.027\,34/2 = 0.013\,67\,\text{m}^3/\text{s}$
and for each drain $Q = ua$.
Substitute: $0.013\,67 = 1.1 \times 0.75(\pi d^2/4)$
from which $d = [(0.013\,67 \times 4)/(1.1 \times 0.75 \times \pi)]^{0.5}$
and $d = 0.145\,\text{m} = 145\,\text{mm}$.

If the nearest standard size of drainpipe is 160 mm, the mean water velocity will be:

$u = Q/a = (0.013\,67 \times 4)/(0.75\pi \times 0.16^2)$

and mean velocity $u = 0.906$ m/s.
Finding the gradient to which each drainpipe must be laid can be done by adopting the Chezy formula $u = C(mi)^{0.5}$
from which $i = u^2/C^2 m$ m/m where for three-quarter flow $m = d/3$
and therefore $dh/L = (0.906^2 \times 3)/(55^2 \times 0.16) = 0.005\,087\,8$
from which $L = 1/0.005\,087\,8 = 197$
and the hydraulic gradient is 1:197.

Summary for Example 7.25
The gradient of 1:197 is the minimum gradient for the drain to achieve a mean water velocity of 0.906 m/s. If the gradient was

Gravitational flow in flooded pipes   161

increased to 1:100 the mean water velocity increases to 1.27 m/s. You should confirm this calculation.

The rainfall intensity of 75 mm/h is not to be considered as acceptable in all design solutions. Rainfall intensities equivalent to 250 mm/h are possible in the United Kingdom although they may last for only a few moments or even seconds. The actual figure selected from Met. data will depend upon from where the run-off is collected.

Example 7.26
An open channel as shown in Figure 7.17 is laid to a gradient of 1:80. If the maximum depth of water flowing is to be 80 mm determine the mass transfer of water in the channel. Take the Chezy coefficient as 55.

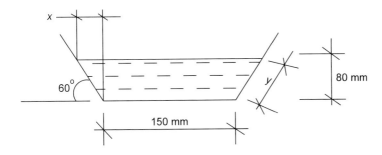

Figure 7.17 Example 7.26.

Solution
The mean hydraulic depth $m =$ (cross-sectional area of flow)/(length of wetted perimeter)

From Figure 7.17 $x = 80 \tan 30 = 80 \times 0.5774 = 46$ mm,
thus cross-sectional area of flow = $(150 + 46) \times 80 = 15\,680$ mm^2 = 0.0157 m^2.

From Figure 7.17 $y = 80/\cos 30 = 92$ mm
thus the wetted perimeter = $150 + 92 + 92 = 334$ mm = 0.334 m
the hydraulic mean depth $m = 0.0157/0.334 = 0.047$.

Adopting the Chezy formula for turbulent flow in open channels:

$$u = C(mi)^{0.5}$$

substituting:

$$u = 55(0.047 \times 1/80)^{0.5} = 1.333 \text{ m/s}$$

and flow rate

$$Q = ua = 1.333 \times 0.08 \times 0.196 = 0.0209 \, \text{m}^3/\text{s}.$$

The mass transfer of water in the channel = 21 kg/s.

7.7 Alternative rational formulae for partial flow

There are three formulae which are adaptions to the Chezy formula and the Darcy equation.

The Manning formula, where

$$u = (1/n)(m^{0.667})(i^{0.5}) \, \text{m/s}$$

where n = the roughness coefficient
= 0.009 for glass smooth pipe
= 0.022 for dirty cast-iron pipe

The roughness coefficient n is not to be confused with the coefficient of friction f.

The Chezy coefficient $C = (2g/f)^{0.5} = 1/n$ and when $C = 55$, $n = 1/55 = 0.0182$
also

$$n = (f/2g)^{0.5}$$

and

$$f = n^2/2g.$$

You should now confirm these relationships.

These relationships beween $n, f,$ and C are tabulated for four values of C and shown in Table 7.3.

Table 7.3 Relationships between the coefficients

n	f	C
0.009	0.0016	111
0.022	0.0095	45.5
0.0182	0.0065	55
0.012	0.0028	84

You should confirm the relationships in Table 7.3.

The Crimp and Bruge's formula took the roughness coefficient n as 0.012 and adapted the Chezy equation thus:

$$u = 84(m^{0.667})(i^{0.5}) \, \text{m/s}.$$

The Darcy–Weisback formula adopts both the Darcy equation for turbulent flow and the Chezy equation, thus:

$$u = (2gmi/f)^{0.5} \, \text{m/s}$$

This formula is in fact the same as the Darcy equation from which the Chezy formula is derived, thus $u = [(2g(d/4)(dh/L)(1/f)]^{0.5}$ m/s from which the Darcy equation $dh = 4fLu^2/2gd$ m of fluid flowing.

You should now confirm that the Darcy–Weisback formula is the same as the Darcy equation. You should also note that the Darcy equation and the Darcy–Weisback formula are given with the hydraulic mean diameter $m = d/4$. This value for m only applies to full bore and half bore flow in circular conduits. A flow of one-third bore in a circular conduit for example, has $m = d/3$ and flow in a channel of rectangular cross-section has $m = db/(2d + b)$. These substitutions must be made before adopting either of these formulae for partial flow.

Example 7.27
Determine the mean fluid velocity of flow in flooded pipes and pipes carrying fluid at half full bore by adopting the following formulae and compare the results. Take the Chezy coefficient as 55 and the hydraulic gradient as 1:50. Chezy formula, Manning formula, Crimp and Bruge's formula, Darcy–Weisback formula.

Solution
The results are tabulated and given in Table 7.4.

You should now confirm the solutions given in Table 7.4. Remember that for flooded pipes and pipes carrying fluid at half full bore $m = d/4$.

Table 7.4 Comparison of fluid velocity, Example 7.27

Source	Mean fluid velocity (m/s)
Chezy	1.23
Manning	0.664
Crimp and Bruge	1.014
Darcy–Weisback	1.23

Note the similarity of solution between Chezy and Darcy–Weisback.

Example 7.28
Determine and compare the mean water velocity and rate of flow in a rectangular channel 150 mm wide and having a water depth of 50 mm when the hydraulic gradient is 1:100. Take the Chezy coefficient as 45.5.

The comparison should be taken using the rational formulae for partial flow.

Solution

The hydraulic mean depth for rectangular channels

$$m = db/(2d + b) = (0.05 \times 0.15)/[(2 \times 0.05) + 0.15] = 0.03.$$

Adopting the Chezy formula $u = C(mi)^{0.5}$

substituting: $u = 45.5[0.03 \times (1/100)]^{0.5} = 0.788\,\text{m/s}$

partial flow $Q = ua = 0.788 \times 0.05 \times 0.15 = 5.91\,\text{l/s}.$

Adopting the Manning formula $u = m^{0.667} \times i^{0.5} \times 1/n$

now $\quad C = (2g/f)^{0.5}$

from which $\quad f = (2g/C^2) = 2 \times 9.81/45.5^2 = 0.0095$

and $\quad n = (f/2g)^{0.5} = (0.0095/2 \times 9.81)^{0.5} = 0.022$

substituting $\quad u = (0.03^{0.667}) \times (1/100)^{0.5} \times (1/0.022)$

from which $\quad u = 0.438\,\text{m/s}$

partial flow $Q = ua = 0.438 \times 0.05 \times 0.15 = 3.29\,\text{l/s}.$

Adopting the Crimp and Bruge's formula $\quad u = 84(m)^{0.667} \times (i)^{0.5}$

substituting: $u = 84 \times 0.03^{0.667} \times (1/100)^{0.5}$

from which $\quad u = 0.81\,\text{m/s}$

partial flow $Q = ua = 0.81 \times 0.05 \times 0.15 = 6.08\,\text{l/s}.$

Adopting the Darcy–Weisback formula $u = (2gmi/f)^{0.5}$
substituting: $u = [2 \times 9.81 \times 0.03 \times (1/100)(1/0.0095)]^{0.5}$
from which $u = 0.787\,\text{m/s}$
partial flow $Q = ua = 0.787 \times 0.05 \times 0.15 = 5.91\,\text{l/s}.$

Summarizing solution to Example 7.28 in Table 7.5.

Table 7.5 Comparison solutions for Example 7.28

Rational formula	Mean velocity	Partial flow
Chezy	0.788 m/s	5.91 l/s
Manning	0.438	3.29
Crimp and Bruge	0.81	6.08
Darcy–Weisback	0.787	5.9

7.8 Chapter closure

You have been introduced to the determination of the flow of fluids considered as non-compressible and subject to the forces of friction in pipes and ducts and to the partial flow of water in pipes, soil stacks and open channels. Successful conclusion of this chapter will enable you to tackle a number of practical problems for which there may be no alternative but to undertake manual solutions.

Natural ventilation in buildings 8

Nomenclature

A	area (m²)
$A_1, A_2,$	free area of openings (m²)
A_3, A_4	free area of openings (m²)
A_s	free area subject to stack effect (m²)
A_w	free area subject to wind (m²)
C	specific heat capacity (kJ/kgK)
C_d	= 0.61, coefficient of discharge
dC_p	difference in pressure coefficient
dP	difference in pressure (Pa)
$d\rho$	difference in air density (kg/m³)
dP_u	difference in velocity pressure (Pa)
dt	difference in temperature (K)
F	force in Newtons
h	height (m)
H	heat energy (kWh)
J	factor for degree of openable window
LMTD	log mean temperature difference (K)
M	mass transfer (kg/s)
N	air change rate per hour
ρ	air density (kg/m³)
ρ_i	air density indoors (kg/m³)
ρ_o	air density outdoors (kg/m³)
P	pressure (Pa)
P_i	pressure of air column indoors (Pa)
P_o	pressure of equivalent air column outdoors (Pa)
P_u	velocity pressure (Pa)
Q_p	plant energy output (kW)
Q_s	volume flow of air due to stack effect (m³/s)
Q_w	volume flow of air due to wind (m³/s)
T_i	absolute indoor temperature (K)
t_i	customary indoor temperature (°C)
t_m	mean temperature (°C)
T_o	absolute outdoor temperature (K)
t_o	customary outdoor temperature (°C)
u	air velocity (m/s)

u_h air velocity at roof level (m/s)
u_m mean air velocity (m/s)
V volume (m³)

The equations with the notation (*) in Sections 8.3, 8.4 and 8.5 are given in the *CIBSE Guide* (1986) section A4.

8.1 Introduction

With the improvement in standards of thermal insulation for the building envelope, the proportion of the plant energy output Q_p required to offset heat loss resulting from natural infiltration of outdoor air has increased.

Modern buildings are better sealed against the random infiltration of outdoor air, but the trend towards the increasing proportion of the building heat loss which has to account for natural infiltration is continuing. Adequate ventilation of the building shell is essential, whereas there is no limitation upon the improvement in the thermal insulation standards of the building envelope.

There are a number of factors which influence the rate of natural ventilation in a building:

wind speed and direction, influenced by geographical location, with respect to the orientation of the building,
the buoyancy forces or stack effect which induces natural draught within the building and depends upon the difference between indoor and outdoor temperature,
the height of the building,
the shape and location of the building with respect to buildings in the vicinity,
wind breaks, natural and artificial,
how well the building is sealed.

The design of the lift shaft, stairwells and atrium, particularly in tall buildings, can have a significant effect upon infiltration initiated by the wind and/or the stack effect.

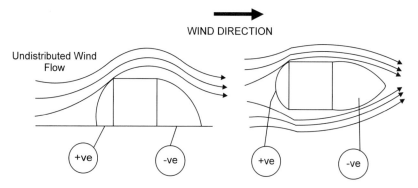

Figure 8.1 Air movement round a building producing +ve and -ve pressures (reproduced with permission of the Heating and Ventilating Contractors' Association).

Aerodynamics around a building 167

As air flows over and around a building it creates positive and negative zones of pressure. Figure 8.1 shows typical wind and pressure patterns in both elevation and plan. Figure 8.2 shows how the prevailing air flow divides over and around a building with the location of the line of maximum air pressure on the facade facing the windward side. Positive pressure (+ve) is created on the windward face and air flow separation occurs at the corners, eaves and roof ridge. Negative pressures (-ve) are generated by air separation along the sides of the building, over the ridge and on the leeward face. Refer to a building in plan and elevation subject to the effects of wind in Figure 8.3. Figure

8.2 Aerodynamics around a building

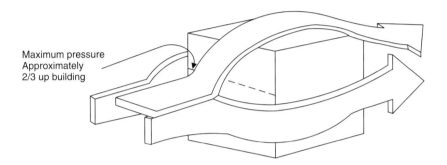

Figure 8.2 Typical air movement over and round a building, one-third over roof, two-thirds round sides (reproduced with permission of the Heating and Ventilating Contractors' Association).

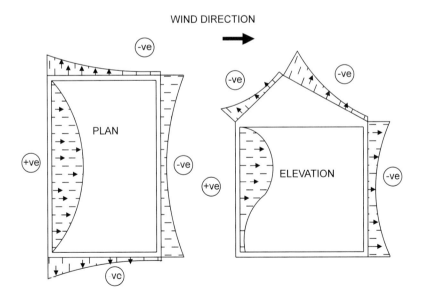

Figure 8.3 Wind pressure distribution on a building (reproduced with permission of the Heating and Ventilating Contractors' Association).

168 Natural ventilation in buildings

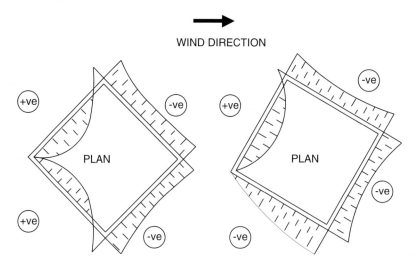

Figure 8.4 Pressure distribution with wind on corner of building (reproduced with permission of the Heating and Ventilating Contractors' Association).

8.4 shows the pressure effects of wind when it is incident on the corner of a building.

The effect of wind pressure on a building will have a significant bearing upon the natural ventilation occurrence. Zones of negative pressure can cause pollution in some rooms on the leeward side from exit points of mechanical extract and from the products of combustion emanating from chimneys. Figure 8.5 shows a building in elevation and the potential zone subject to the effects of pollution from the point

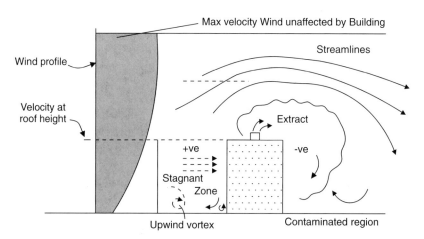

Figure 8.5 Flow patterns over a building showing effect of building height and pollution emission (reproduced with permission of the Heating and Ventilating Contractors' Association).

Aerodynamics around a building 169

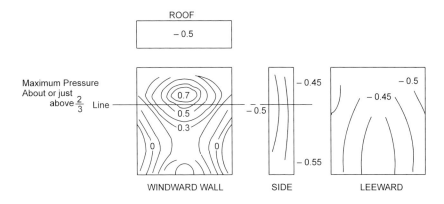

Figure 8.6 Typical pressure distribution for an average building form. (Greatest pressure is at two-thirds point on windward wall.) (Reproduced with permission of the Heating and Ventilating Contractors' Association.)

of extract. The wind velocity profile shown in Figure 8.5 will vary with the roughness of the underlying surfaces, or terrain. As indicated in Figure 8.2 the area of maximum wind pressure occurs at around 2/3 of the height of the building. Figure 8.6 shows a typical pressure distribution on a vertical wall facing the wind.

Invariably a building is not located in isolation but forms part of a group of buildings which can vary in density and relative position. Figure 8.7 shows wind flow patterns around a group of buildings and Figure 8.8 shows the wind flow patterns over the same group of buildings. The zones of positive and negative pressures are identified as well as the potential zone where pollution may be a cause for concern.

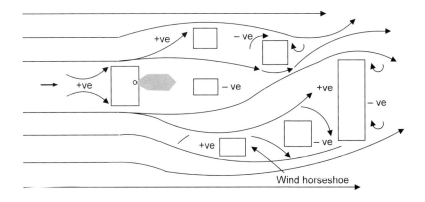

Figure 8.7 Plan of building complex (reproduced with permission of the Heating and Ventilating Contractors' Association).

170 Natural ventilation in buildings

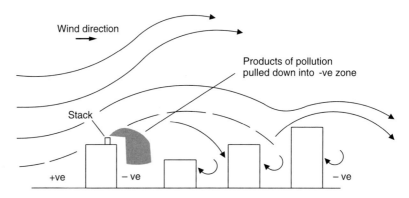

Figure 8.8 Elevation of building complex (reproduced with permission of the Heating and Ventilating Contractors' Association).

The wind flow patterns and the effects that groups of buildings have in the path of the wind can be developed using models in a wind-tunnel. The factors which influence the effects of wind are:

- building shape, size and orientation
- location of the building with respect to other properties, including their shapes
- natural and artificial wind-breaks
- type of terrain
- wind speed and direction
- height above sea level.

It is apparent from the foregoing that the effects of wind on the natural ventilation occurrence in buildings is a complex subject and one which requires the use of the wind-tunnel and computer modelling techniques. However the magnitude and characteristics of natural ventilation can be demonstrated by considering simplified models and by adopting empirical equations.

8.3 Effects on cross ventilation from the wind

Wind pressure which results from its velocity can be obtained from the velocity pressure term of the Bernoulli equation referred to in Chapter 7:

thus $P_u = 0.5\rho u^2$ Pa.

If the initial wind velocity is u and the final wind velocity is zero

$dP_u = 0.5\rho u^2$ Pa.

Example 8.1
Determine the pressure caused by the following wind speeds on the facade of a building: 20 km/h, 40 km/h, 80 km/h. Take air density as 1.2 kg/m^3.

Solution
Air speed in m/s $= 20 \times 1000/3600 = 5.556$ m/s, 11.11 m/s and 22.22 m/s.

$$dP_u = 0.5 \times 1.2 \times 5.556^2 = 18.52 \text{ Pa}, 74.1 \text{ Pa}, \text{ and } 296 \text{ Pa}.$$

You will notice that since $dP_u \propto u^2$, as wind speed doubles so velocity pressure quadruples. It is also important to appreciate that for example $296 \text{ Pa} = 296 \text{ N/m}^2$ of facade and although the pressure on the facade of a building is not constant (Figure 8.6), the gross lateral force F in Newtons on a building facade measuring 10 m by 15 m high and subject to a wind speed of 80 km/h (50 mph) will be:

$$F = P \times A = 296(10 \times 15) = 44\,400 \text{ N}.$$

This is equivalent to a gross lateral load of $44\,400/9.81 = 4526$ kg or 4.526 t.

Air flow through openings

The rate of air flow subject to wind through an opening is expressed as:

$$Q_w = A_w C_d (2dP/\rho)^{0.5} \text{ m}^3/\text{s} \tag{*}$$

or $\quad Q_w = A_w C_d u_m (dC_p)^{0.5} \text{ m}^3/\text{s}. \tag{*}$

If the openings are in series as shown in Figure 8.9

$$1/A_w^2 = [1/(A_1 + A_2)^2] + [1/(A_3 + A_4)^2] \tag{*}$$

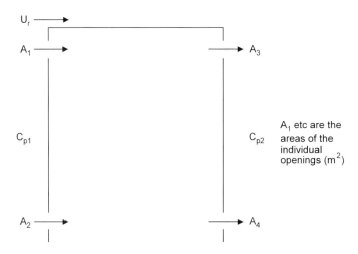

Figure 8.9 Cross ventilation of a simple building due to wind forces only (reproduced with permission of the Heating and Ventilating Contractors' Association).

Example 8.2
(a) Determine the minimum ventilation rate in a workshop due to a mean wind speed of 9 m/s on one of its facades in which there are two openings. The lower opening has a free area of 0.8 m² and the upper opening a free area of 0.3 m². Assume that there are similar openings on the opposite facade.

Take air density as 1.2 kg/m^3, $dC_p = 1.0$ and $C_d = 0.61$.

(b) If the workshop measures 50 m × 28 m × 5 m high determine the infiltration rate of air resulting from the wind.

Solution (a)
For openings in series $1/A_w^2 = [1/(0.8 + 0.3)^2 + 1/(0.8 + 0.3)^2]$

from which $\quad 1/A_w^2 = 0.8264 + 0.8264 = 1.653$

and $\quad\quad\quad A_w^2 = 0.605$

therefore $\quad\quad A_w = 0.778 \text{ m}^2$

and $\quad\quad\quad Q_w = A_w C_d u_m (dC_p)^{0.5}$

If $dC_p = 1.0$ and $C_d = 0.61$, $Q_w = 0.778 \times 0.61 \times 9 \times 1.0$ and $Q_w = 4.27 \text{ m}^3/\text{s}$.

Now adopting the equation $Q_w = A_w C_d (2dP/\rho)^{0.5}$.

If the pressure drop dP across the building is taken as the drop in velocity pressure dP_u where final velocity is taken as zero then $dP = dP_u$

thus

$$dP = dP_u = 0.5\rho u^2 = 0.5 \times 1.2 \times 9^2 = 48.6 \text{ Pa}.$$

Substituting:

$$Q_w = 0.778 \times 0.61 \times (2 \times 48.6/1.2)^{0.5}.$$

therefore $\quad Q_w = 4.27 \text{ m}^3/\text{s}$.

Solution (b)
Air change rate N can be obtained from $Q_w = NV/3600 \text{ m}^3/\text{s}$

rearranging $\quad N = 3600 Q_w/V$

$N = 3600 \times 4.27/(50 \times 28 \times 5)$

$N = 2.2$ air changes per hour

Summary for Example 8.2
The wind speed is quite high; 9 m/s is equivalent to 20 mph. This is the reason for the high rate of air change. Note that no account has

been taken of natural ventilation due to temperature difference between indoors and outdoors.

If the wind speed is reduced to 3 m/s which is a more normal value for a less exposed site, the volume flow rate Q_w attributable to wind speed will be 1.424 m³/s and the air change rate N will be 0.732 per hour. You should now confirm these solutions.

8.4 Stack effect

The difference in temperature between inside a building and outside creates thermal forces called stack effect. The more extreme the temperature difference the greater is the potential for outdoor air to enter the building thus forcing the warm air inside, outside. The resulting stack effect is caused by the difference in density between indoor air and air outdoors and the effect is most noticeable during the winter when the greatest temperature difference will be apparent for a heated building.

Figure 8.10 illustrates the stack effect by showing the column of warm air inside the building and a corresponding column of cold air outside. The pressure at the base of the column of outdoor air will be $P_o = h\rho_o g$ Pa and the pressure at the base of the column of warm air indoors $P_i = h\rho_i g$ Pa. The air densities are taken as mean values over the height of the column h. The pressure difference dP provides the driving force for air movement from indoors to outdoors.

Thus $dP = h(d\rho)g$ Pa.

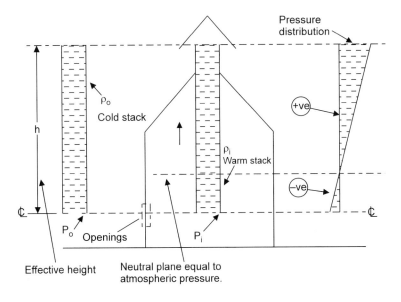

Figure 8.10 Thermal forces. (Stack effect.) (Reproduced with permission of the Heating and Ventilating Contractors' Association.)

Figure 8.10 also shows the pressure distribution such that at low level the air movement into the building is subject to suction or negative pressure (-ve) due to its buoyancy and the exit point of the air from the building at high level is subject to positive pressure (+ve). The neutral point occurs at a horizontal plane where the negative air pressure changes to positive pressure. The neutral point is at atmospheric pressure.

There are two methodologies for the determination of the stack effect.

1. Air density at 20°C and 101 325 Pa is 1.2 kg/m^3, and frequently this is taken as the mean density of the air indoors. Since air density is inversely proportional to its absolute temperature and atmospheric pressure is considered constant both indoors and outdoors, the density of outdoor air can then be obtained from $\rho_o = \rho_i(T_i/T_o)$

   from which $dP = h(d\rho)g$ Pa.

   Alternatively air densities can be obtained from the *Thermodynamic and Transport Properties of Fluids*.

2. However it is more convenient to determine the pressure drop caused by the stack effect from a knowledge of the mean temperature of each column of air.

From the equation $dP = h(d\rho)g$, the pressure difference can be expressed as:

$dP = hg(\rho_o - \rho_i)$ Pa.

The density of outdoor air at 0°C, 273 K is 1.293 kg/m^3

This air density can be put in the form $\rho_o = (1.293 \times 273)/(273 + t_o)$.

The density of the air indoors will therefore be

$\rho_i = (1.293 \times 273)/(273 + t_i)$.

Substituting these two equations into the formula for dP:

$dP = hg[(1.293 \times 273)/(273 + t_o) - (1.293 \times 273)/(273 + t_i)]$

thus $dP = (1.293 \times 273)hg[(1/(273 + t_o)) - (1/(273 + t_i))]$

from which $dP = 3463h[(1/(273 + t_o)) - (1/(273 + t_i))]$. (*)

Example 8.3
A building is 15 storeys high and held at a temperature of 20°C. Determine the potential stack effect when outdoor temperature is −4°C given that the floor to ceiling height is 3 m.

Solution
Assuming that the stack effect extends to the full height of the building

$dP = 3463 \times 15 \times 3[(1/268) - (1/293)] = 49.6\,\text{Pa}$.

Now try $dP = h(d\rho)g\,\text{Pa}$

Cross ventilation through openings
The rate of air flow subject to stack effect through an opening is:

$$Q_s = C_d A_s [(2(dt)hg)/(t_m + 273)]^{0.5}\,\text{m}^3/\text{s} \qquad (*)$$

where for apertures in series as shown in Figure 8.11:

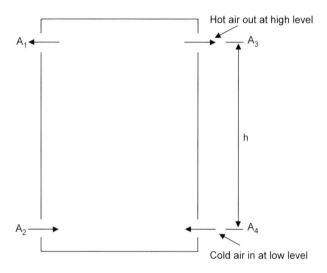

Figure 8.11 Cross ventilation of a simple building due to temperature difference only (reproduced with permission of the Heating and Ventilating Contractors' Association).

$$1/A_s^2 = [(1/(A_1 + A_3))^2 + (1/(A_2 + A_4))^2]. \qquad (*)$$

Note the difference in this equation for A_s compared with the equation for A_w.

Example 8.4
Determine the ventilation rate due to stack effect for the building in Example 8.3 given that the lower and upper openings on one facade are 0.5 and 0.3 m² with similar openings on the facade opposite to it. Take C_d as 0.61.

Solution

Substituting : $1/A_s = [(1/(0.8)^2) + (1/(0.8^2))]$

from which $A_s = 0.32\,\text{m}^2$

thus $Q_s = 0.61 \times 0.32[(2(20+4)(3 \times 15) \times 9.81)/(12+273)]^{0.5}$

$Q_s = 1.683\,\text{m}^3/\text{s}.$

Summary for Examples 8.2 and 8.4
These examples deal with uninhibited cross ventilation where there are no internal partitions. For these simple applications the actual ventilation rate may be taken as the larger of that due to the wind or stack effect. It is likely that in the summer the building will be subject to moderate natural ventilation since wind speed will normally be low with a small difference between indoor and outdoor temperature. If the building is air conditioned with indoor temperature lower than outdoor temperature at times during the summer, the stack effect will be reversed. This means that the cooler air from the building will emanate from openings at low level.

8.5 Natural ventilation to internal spaces with openings in one wall only

Figure 8.12 refers to the effect of wind incident upon a facade having one opening where the approximate volume flow $Q_w = 0.025Au_h\,\text{m}^3/\text{s}.(*)$ The velocity of the wind tends to increase with height above ground level (Figure 8.5).

Figure 8.13 refers to the effect of indoor to outdoor temperature difference for one opening in the facade where the approximate volume flow

$$Q_s = C_d(A/3)J[((dt)hg)/(t_m + 273)]^{0.5}\,\text{m}^3/\text{s}. \qquad (*)$$

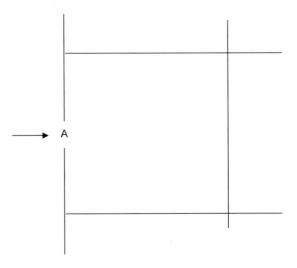

Figure 8.12 Internal space subject to wind.

Natural ventilation to internal spaces with openings in one wall only 177

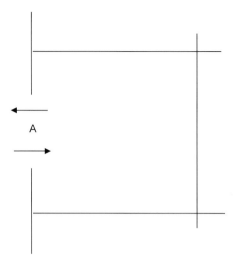

Figure 8.13 Internal space subject to temperature difference.

Table 8.1 Values of J for two types of window. (Reproduced from *CIBSE Guide* section A4 (1986) by permission of the Chartered Institution of Building Services Engineers.)

Angle of opening	Type of window	J
30	side mounted casement	0.6
60	side mounted casement	0.9
90	side mounted casement	1.1
30	centre pivoted window	0.7
60	centre pivoted window	0.92
90	centre pivoted window	0.95

The value of J depends upon the angle of opening for the window. Table 8.1 lists some typical values.

Example 8.5
Determine the volume flow rate of outdoor air into an internal space having an opening on the windward side whose equivalent area is 6400 mm². The mean wind speed at the building height of 10 m is 9 m/s.

Solution
Substituting into the formula we have

$$Q_w = 0.025 \times (6400/1\,000\,000) \times 9 = 0.00144 \, \text{m}^3/\text{s}$$
$$Q_w = 1.44 \, \text{l/s}.$$

Example 8.6

(a Determine the volume flow rate of air exchange through a side-mounted casement window to an internal space given that the angle of opening is 60°. Indoor temperature is 25°C, outdoor temperature is 10°C and the height of the casement is 0.8 m by 0.5 m wide.

(b Determine the air change rate for the room as a consequence of the opened window given that it measures $5 \times 4 \times 2.7$ m high.

Solution (a)

The value of J, from Table 8.1 is 0.9 and $t_m = (25 + 10)/2 = 12.5°C$.

Substituting the data into the equation for Q_s for natural ventilation due to temperature difference:

$$Q_s = (0.61 \times (0.8 \times 0.5)/3) \times 0.9[(25 - 10) \times 0.8 \times 9.81)/(12.5 + 273)]^{0.5}$$

$$Q_s = 0.047 \, m^3/s$$

$$Q_s = 47 \, l/s.$$

Solution (b)

From the equation $Q = NV/3600 \, m^3/s$,

the air change rate $N = 3600Q/V = 3600 \times 0.047/(5 \times 4 \times 2.7)$

$$= 3.13 \text{ per hour}.$$

8.6 Ventilation for cooling purposes

Current design favours the use of natural or fan-assisted ventilation of the building shell for maintaining comfort conditions in preference to the use of air conditioning plant. The building envelope must be designed and orientated in order to take advantage of the wind and the stack effect caused by indoor to outdoor differences in temperature. A number of articles relating to buildings designed in this manner have appeared in *Building Services*, the monthly journal of the Chartered Institution of Building Services Engineers.

Where air conditioning plant is required and even in place of it, advantage can be made for night-time cooling of the building shell using natural ventilation. Again the design of the building must account for the flow paths for the ventilating outdoor air to ensure cooling of the building structure and exfiltration of the resultant warmed air. This involves analysis of the building's thermal response to summer outdoor temperatures and solar heat gains to ensure a

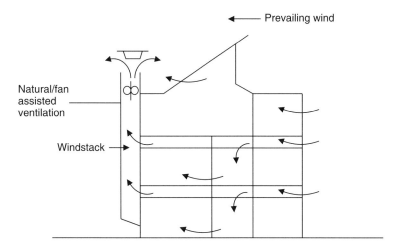

Figure 8.14 Section through a building designed for cooling by ventilation.

thermally stabilized building structure at the commencement of occupation at the beginning of the day.

Figure 8.14 shows a section through a building designed for cooling by ventilation.

Case study 8.1

A five-storey building measures 30×15 m and has floor to ceiling heights of 3 m. It is to be structurally cooled in the summer months during the evening and night by natural ventilation to provide a total of 42 air changes during the unoccupied period which extends from 1800 hours to 0800 hours.

Determine the free area of each of the two openings on the leeward side of the building:

(a) given design mean indoor and outdoor air temperatures during the unoccupied period are 25°C and 15°C respectively.
(b) for a wind speed of 5 m/s.

Data: The two openings on the windward side of the building have free area of $1.5\,\text{m}^2$ at low level and $1.8\,\text{m}^2$ at high level. Take $C_d = 0.61, h = 13.5$ m.

Assume the openings are in series.

(c) Estimate the rate of free cooling from the natural ventilation of the building when indoor temperature is 25°C and outdoor temperature is 15°C.
(d) Estimate the daily cooling energy extraction by natural ventilation between the hours of 1800 and 0800.

SOLUTION (A)

The required air change rate per hour = 42/(1800 hours to 0800 hours) = 42/14 = 3

From $Q = NV/3600$

$$Q = 3 \times (30 \times 15 \times 3 \times 5)/3600 = 5.625 \, \text{m}^3/\text{s}.$$

Estimating for temperature difference

$$Q_s = C_d A_s [(2(dt)hg)/(t_m + 273)]^{0.5} \, \text{m}^3/\text{s}$$

where $t_m = (15 + 25)/2 = 20°C$

Substituting: $5.625 = 0.61 A_s [(2(25 - 15) \times 13.5 \times 9.81)/(20 + 273)]^{0.5}$

from which $5.625 = 0.61 A_s \times 3.0066$

and $A_s = 3.07 \, \text{m}^2$.

For apertures in series $1/A_s^2 = [(1/(A_1 + A_3)^2) + (1/(A_2 + A_4)^2]$

Substituting: $1/3.07^2 = [(1/(1.8 + A_3)^2) + (1/(1.5 + A_4)^2]$.

Assuming $A_3 = A_4 = A$

$$0.1061 = [(1/1.8 + A)^2) + (1/(1.5 + A)^2].$$

To solve the equation it can be expressed as $z = x + y$ where $z = 0.1061$ and with values allocated for A, the free area of openings A_3 and A_4.

The solution is tabulated and given in Table 8.2.

Table 8.2 Solution to openings for free areas A_3 and A_4, case study 8.1a

A	x	y	z
2	0.06925	0.8163	0.1509
2.5	0.05408	0.0625	0.1166
2.7	0.04938	0.05669	0.10607

From Table 8.2 the free area for apertures A_3 and A_4 is approximately $2.7 \, \text{m}^2$ each since z is almost equal to 0.1061.

SOLUTION (B)

For ventilation resulting from wind $Q_w = C_d A_w (2dP/\rho)^{0.5} \, \text{m}^3/\text{s}$

now $dP = dP_u = 0.5 \rho u^2$.

At a mean temperature $t_m = 20°C$, $\rho = 1.2 \, \text{kg/m}^3$

thus $dP = 0.5 \times 1.2 \times 5^2 = 15 \, \text{Pa}$

and $Q_w = 0.61 A_w (2 \times 15/1.2)^{0.5} = 3.05 A_w$

substituting for Q_w: $5.625 = 3.05 A_w$

from which $A_w = 1.844 \, \text{m}^2$.

For apertures in series $1/A_w^2 = [(1/(A_1 + A_2)^2) + (1/(A_3 + A_4)^2)]$

substituting: $1/1.844^2 = [(1/1.5 + 1.8)^2) + (1/(A_3 + A_4)^2)]$.

If $A_3 = A_4$ $1/3.4 = [(1/10.89) + (1/(2A))^2]$

rearranging $0.294 - 0.0918 = 1/(2A)^2$

from which $0.2022 = 1/(2A)^2$

and $4.946 = (2A)^2$

thus $2.224 = 2A$

and $A = 1.112 \, \text{m}^2$

Therefore the free area of each of apertures A_3 and A_4 is $1.112 \, \text{m}^2$.

SOLUTION (C)

The building structure will absorb heat by solar radiation and conduction during the hours of daylight. With the ventilation system in use during occupation this will help to ensure against excessive indoor temperatures.

The mass transfer of air through the building by natural ventilation

$M = Q\rho = 5.625 \times 1.2 = 6.75 \, \text{kg/s}$.

The maximum rate of cooling between 1800 hours and 0800 hours can be determined from:

rate of cooling $= MCdt \, \text{kW}$

where for air $C = 1.025 \, \text{kJ/kgK}$

substituting: the rate of free cooling $= 6.75 \times 1.025 \times (25 - 15) = 69 \, \text{kW}$.

SOLUTION (D)

The estimated daily cooling energy extracted by natural ventilation can be determined by taking log mean temperature difference between indoors and outdoors between 1800 hours and 0800 hours. Log mean temperature difference accounts for there being a change both in outdoor temperature and a change in indoor temperature and the need to find the true mean difference between these two temperature changes.

At the beginning of the unoccupied period when cooling of the building is considered at 1800 hours, it is possible that indoor temperature and outdoor temperature will be equal at 25°C and no cooling takes place. At some point during the night outdoor temperature drops to 15°C and with indoor temperature still at or near 25°C, the maximum cooling rate will be 69 kW and the maximum temperature difference is $(25 - 15) = 10$ K. The minimum temperature difference will occur at 0800 hours when indoor temperature will be about 18°C and outdoor temperature 15°C. Thus minimum temperature difference $(18 - 15) = 3$ K. The log mean temperature difference will be

$$\text{LMTD} = dt_{max} - dt_{min})/\ln(dt_{max}/dt_{min})$$

thus LMTD $= (10 - 3)/\ln(10/3) = 5.814$ K.

Note: the *arithmetic* (less accurate) mean temperature difference $= (10 + 3)/2 = 6.5$ K.

Now the daily energy extracted $H = MCdt \times$ time kWh.

The units of the terms are: $H = $ (kg/s)(kJ/kgK)(K)(hours)
$\qquad\qquad\qquad\qquad\quad = $ (kJ/s)(hours) $=$ kWh

substituting: daily energy extracted $= 6.75 \times 1.025 \times 5.814$
$\qquad\qquad\qquad\qquad\qquad\qquad\quad \times (1800 - 0800)$
$\qquad\qquad\qquad\qquad\qquad\qquad = 563$ kWh.

SUMMARIZING THE SOLUTIONS TO CASE STUDY 8.1

Maximum rate of cooling by natural ventilation at night $= 69$ kW.

Estimated daily cooling energy from natural ventilation $= 563$ kWh.

Conditions	A_1	A_2	A_3	A_4
Stack effect, $t_i = 25°C$, $t_o = 15°C$	1.8 m²	1.5 m²	2.7 m²	2.7 m²
Wind effect, $u = 5$ m/s	1.8	1.5	1.112	1.112

There is a substantial difference in free area for apertures A_3 and A_4 between the stack effect and the effect of the wind. You should now consider the effect of a wind speed of 3 m/s upon size of apertures A_3 and A_4 as air movement at night can be quite low in the summer. The prevailing conditions will be a combination of wind and stack effect. However the larger apertures determined from either wind or temperature difference would be considered appropriate.

QUALIFYING REMARKS RELATING TO CASE STUDY 8.1

There have been a number of assumptions made in the solution to case study 8.1 and the following qualifying remarks must be made.

- It is assumed cross-ventilation takes place with no internal partitions.
- Recourse should be made to establish minimum summertime outdoor temperatures which normally occur at night-time. This will depend upon geographical location.
- The building's thermal capacity and orientation including the ventilation pathways need analysing to ensure that peak indoor temperature normally occurs at the end of, or after, the occupation period of 1800 hours.
- Peak indoor temperature will also need to be set as a design parameter in the modelling process.
- Four apertures in series have been considered, two on the windward and two on the leeward side of the offices.

8.7 Fan assisted ventilation

If the building is located in a sheltered position, the use of extract fans can provide a positive air displacement for the building. They can also be used to advantage when the indoor to outdoor temperature difference is small thus reducing the influence of the stack effect. The extract fans can be controlled by wind speed and direction and indoor temperature so that they are only used when necessary to aid in capturing the heat energy absorbed by the building during the day. Refer again to Figure 8.14.

8.8 Chapter closure

You now have knowledge of the forces and factors affecting the natural ventilation of buildings with respect to stack effect and wind. You have the skills required to undertake simple modelling processes relating to the size and location of apertures in the building envelope and to the mass transfer of air through the building. An approximate methodology to estimate the cooling effect of night-time ventilation of the building structure has been investigated.

For realistic assessment of natural ventilation as a means of structurally cooling a building at night, recourse must be made to computerized modelling techniques.

9 Regimes of fluid flow in heat exchangers

Nomenclature

A	heat exchange surface (m²)
C_c	specific heat capacity of cold fluid (kJ/kgK)
C_h	specific heat capacity of hot fluid (kJ/kgK)
CR	capacity ratio
dt	temperature difference (K)
dt_m	true mean temperature difference (K)
dt_{max}	maximum temperature difference (K)
dt_{min}	minimum temperature difference (K)
E	effectiveness
exp	exponential
f	correction for cross flow
h	specific enthalpy of the superheated vapour (kJ/kg)
h_f	specific enthalpy of the saturated liquid (kJ/kg)
h_{fg}	latent heat of evaporation (kJ/kg)
h_{si}	inside heat transfer coefficient (kW/m²K)
h_{so}	outside heat transfer coefficient (kW/m²K)
HVAC	heating ventilation and air conditioning
h_w	specific enthalpy of the wet vapour (kJ/kg)
k	thermal conductivity (kW/mK)
L	length (m)
LMTD	log mean temperature difference (K)
M_c	mass flow of cold fluid (kg/s)
M_h	mass flow of hot fluid (kg/s)
NTU	number of transfer units
Q	output (kW)
q	dryness fraction
r_1, r_2	radius for radial heat transfer (m)
Re	Reynolds number
R_f	fouling thermal resistance (m²K/kW)
R_t	total thermal resistance (m²K/kW)
t_{c1}, t_{c2}	inlet and outlet temperatures of the secondary fluid (°C)
t_{h1}, t_{h2}	inlet and outlet temperatures of the primary fluid (°C)
t_m	mean temperature (K)
U	overall heat transfer coefficient (kW/m²)
U_L	overall heat transfer coefficient (kW/mK)
Z_1, Z_2	temperature ratios for cross flow

9.1 Introduction

There are many different types of heat exchanger available to the building services industry. Plate heat exchangers and heat pipes which have been in use in other industries for many years are now used in HVAC systems for extracting low-grade heat from return air in ventilation and air conditioning systems, for example. The thermal wheel is also used for this purpose. The cooling tower employed for cooling condenser water in an air conditioning plant allows the two fluids, atmospheric air and condenser water, to come into direct contact for heat transfer to take place.

This chapter focuses on heat exchangers having a solid boundary between the two fluids. The function of this type of heat exchanger is to allow the transfer of heat energy between two fluids at different temperatures across the solid boundary. It is used to ensure that the two fluids do not come into direct contact.

An ideal heat exchanger of this type should achieve maximum rate of heat exchange using the minimum heat exchange space and the minimum pressure drop on both sides of the solid boundary. In practice if the solid boundary is a plain straight tube a comparatively large heat exchange space will be required although the pressure drop will be relatively low. Alternatively a solid boundary in the form of a coiled tube with finning extends the heat transfer surface in a small space but the pressure loss inside and outside the tube bundle will be comparatively high.

Inevitably a compromise is usually made in heat exchanger design for specific applications. Table 9.1 lists some of the heat exchangers used in the building services industry. You should familiarize yourself with the construction of the various heat exchangers on the market from manufacturers' current literature.

Table 9.1 Examples of heat exchangers in the building services industry

Heat exchanger	Media
Double pipe	Water
Shell and tube	Water/condensing/evaporating fluids
Plate	Water/air, water/water
Run around coils	Water/air
Pipe coils	Water/air
Heat pipes	Water/air
Regenerator (thermal wheel)	Air/air
Spray condenser/desuperheater/flash steam recovery/cooling tower	Water/air, condense/steam, air/water

9.2 Parallel flow and counterflow heat exchangers

Figure 9.1 shows a parallel flow heat exchanger with its accompanying temperature distribution assuming both fluids vary in temperature. t_{h1} being the initial temperature of the hot fluid and t_{c1} the initial temperature of the cold fluid. Figure 9.2 shows a heat exchanger in

186 Regimes of fluid flow in heat exchangers

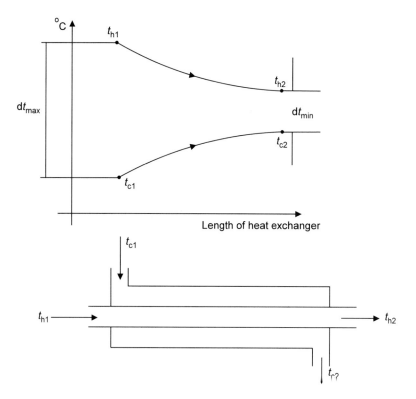

Figure 9.1 Parallel flow.

counterflow with its accompanying temperature distribution, assuming both fluids vary in temperature on their passage through the heat exchanger. Again t_{h1} is the initial temperature of the hot fluid and t_{c1} the initial temperature of the cold fluid.

For practical reasons, for a heat exchanger in parallel flow the final temperatures t_{h2} and t_{c2} can never be equal and clearly t_{c2} cannot exceed t_{h2}. However, for a heat exchanger in counterflow t_{c2} can exceed t_{h2}. Refer to Figure 9.2. Therefore parallel flow has a limitation on the relationship between the primary and secondary leaving temperatures.

In cases where both the primary and secondary fluids vary in temperature the arithmetic mean temperature difference which provides the motive force in the heat exchange does not always register the true mean temperature difference and the log mean temperature difference between the two fluids is adopted.

$$\text{LMTD } dt_m = (dt_{max} - dt_{min})/(\ln(dt_{max}/dt_{min})) \text{ K}$$

If however $dt_{max} = dt_{min}$, LMTD dt_m = zero and the arithmetic mean temperature difference is used.

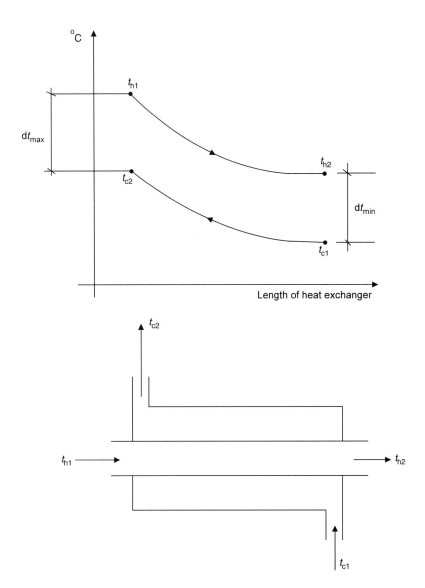

Figure 9.2 Counterflow.

For heat exchangers in parallel flow primary temperatures $t_{h1} \rightarrow t_{h2}$
secondary temperatures $t_{c1} \rightarrow t_{c2}$
from which $dt_{max} = t_{h1} - t_{c1}$ and $dt_{min} = t_{h2} - t_{c2}$.
For heat exchangers in counterflow primary temperatures $t_{h1} \rightarrow t_{h2}$
secondary temperatures $t_{c2} \rightarrow t_{c1}$
from which $dt_{max} = t_{h1} - t_{c2}$ and $dt_{min} = t_{h2} - t_{c1}$.
For both types of flow dt_{max} and dt_{min} can be reversed.

188 Regimes of fluid flow in heat exchangers

Example 9.1
Determine the true temperature difference for the primary and secondary fluids for a heat exchanger in counterflow.

(a) Primary fluid inlet temperature 120°C outlet temperature 90°C. Secondary fluid inlet temperature 10°C outlet temperature 80°C.
(b) Primary fluid 100°C inlet temperature 80°C outlet temperature. Secondary fluid 10°C inlet temperature 30°C outlet temperature.

Solution
(a) Primary temperatures $\quad 120 \longrightarrow 90$
$\quad$ Secondary temperatures $\quad \underline{80 \longleftarrow 10}$
$\quad t_{max,min} \qquad\qquad\qquad dt_{min}\,40 - dt_{max}\,80$

and LMTD $dt_m = (80 - 40)/\ln(80/40) = 57.7\,\text{K}$

(b) Primary temperatures $\quad 100 \longrightarrow 80$
$\quad$ Secondary temperatures $\quad \underline{30 \longleftarrow 10}$
$\quad t_{max,min} \qquad\qquad\qquad dt_{min}\,70 - dt_{max}\,70$

and LMTD $dt_m = $ zero.

The arithmetic mean temperature difference must therefore be used here and will be:

Primary mean temperature $(100 + 80)/2 = 90$
Secondary mean temperature $(30 + 10)/2 = 20$
Arithmetic mean temperature difference $dt_m = 90 - 20 = 70\,\text{K}$.

BOILING AND CONDENSING IN PARALLEL AND COUNTERFLOW

It is possible for one of the fluids to remain at constant temperature during the process of heat exchange across the solid boundary. The evaporator is an example in which the fluid being cooled causes the cool fluid to evaporate at constant temperature. Figure 9.3 shows the heat exchanger and the accompanying process.

The condenser is another example in which the fluid being condensed at constant temperature causes the cool fluid to rise in temperature. Figure 9.4 shows the heat exchanger and the accompanying process.

9.3 Heat transfer equations

The rate of heat transfer may be expressed in various ways:

$$Q = UA\,dt_m \text{ kW}$$

where $U = 1/R_t \text{ kW/m}^2\text{K}$

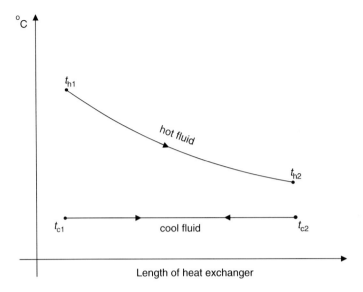

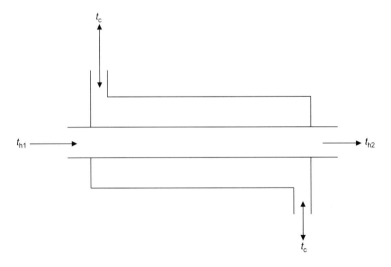

Figure 9.3 Boiling in a shell and tube exchanger, cool fluid boiling.

and $R_t = (1/h_{si}) + R_f + (1/h_{so})\, \text{m}^2\text{K/kW}$.

h_{si} and h_{so} derive from the laminar sublayers either side of the solid boundary, see Chapter 6. The thermal resistance of the solid boundary is not significant and therefore sometimes ignored.

$$Q = M_c C_c (t_{c2} - t_{c1})\, \text{kW}$$

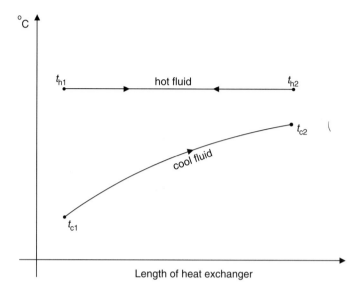

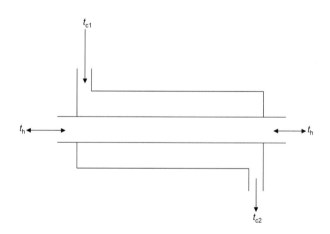

Figure 9.4 Condensing in a shell and tube exchanger, hot fluid condensing.

$$Q = M_h C_h (t_{h1} - t_{h2}) \text{ kW}$$
$$Q = M_h h_{fg} = M_h (q h_{fg}) \text{ kW}$$
$$Q = M_h (h - h_f) = M_h (h_w - h_f) \text{ kW}.$$

Ignoring the inefficiency of heat exchange a heat balance may be drawn such that: heat lost by the primary fluid = heat gained by the secondary fluid.

For example:

$$M_h C_h(t_{h1} - t_{h2}) = M_c C_c(t_{c2} - t_{c1})$$

from which one unknown can be evaluated.

FOULING FACTORS

The overall heat transfer coefficient U for the solid boundary between the primary and secondary fluids, introduced above, should account for a layer of dirt or scale on the heat exchanger surface in contact with the fluid. Expressed as a thermal resistance R_f to heat flow, it will be at a minimum value at commissioning and reach a maximum resistance at the point when cleaning and descaling is scheduled. In practice it is difficult to evaluate and depends upon:

fluid properties
fluid temperatures
fluid velocities
heat exchanger material and materials used elsewhere in the system
heat exchanger configuration.

Two approaches to accounting for fouling resistance include making a correction to the overall heat transfer coefficient for the heat exchanger or alternatively calculating the fouling resistance from various trials thus:

$$R_f = (1/U\text{dirty}) - (1/U\text{clean})$$

Clearly the effects of scale and dirt, if not dealt with under a regulated planned maintenance, may have a significant effect upon the performance of the heat exchanger. Key issues would include the need for water treatment, inspection, cleaning and flushing out.

There now follows two examples relating to the subject matter set out above.

Example 9.2
A hot fluid is cooled from 118°C to 107°C in a double pipe heat exchanger. Assuming the overall heat transfer coefficient to remain constant, compare the advantage of counterflow over parallel flow in the amount of heat transfer area required when a cold fluid is to be heated from:

(a) 57°C to 104°C.
(b) 30°C to 77°C.
(c) 10°C to 57°C.

Solution
You should note that the temperature rise of the secondary fluid is 47 K in each case. Examples of double pipe heat exchangers are shown in Figures 9.1 and 9.2.
Consider counterflow:

	(a)		(b)		(c)	
primary fluid	118 $\rightarrow$ 107		118 $\rightarrow$ 107		118 $\rightarrow$ 107	
secondary fluid	104 $\leftarrow$ 57		77 $\leftarrow$ 30		57 $\leftarrow$ 10	
$dt_{max,min}$	14	50	41	77	61	97

LMTD $dt_m = (50 - 14)/\ln(50/14) = 28.3\,K$
$dt_m = (77 - 41)/\ln(77/41) = 57.1\,K$
$dt_m = (97 - 61)/\ln(97/61) = 77.6\,K$

Consider parallel flow:

	(a)		(b)		(c)	
primary fluid	118 $\rightarrow$ 107		118 $\rightarrow$ 107		118 $\rightarrow$ 107	
secondary fluid	57 $\rightarrow$ 104		30 $\rightarrow$ 77		10 $\rightarrow$ 57	
$dt_{max,min}$	61	3	88	30	108	50

LMTD $dt_m = (61 - 3)/\ln(61/3) = 19.3\,K$
$dt_m = (88 - 30/\ln(88/30) = 53.9\,K$
$dt_m = (108 - 50)/\ln(108/50) = 75.3\,K$

Since the overall heat transfer coefficient remains constant the true mean temperature differences can now be used to show the advantage of counterflow over parallel flow in relation to the amount of heat transfer surface required.

(a) $28.3/19.3 = 1.466$, (b) $57.1/53.9 = 1.059$, (c) $77.6/75.3 = 1.031$.

Analysis of Example 9.2
Although the secondary fluid has the same temperature rise in each case the advantage of counterflow diminishes as the inlet temperature of the secondary fluid is reduced. Thus as $(t_{h2} - t_{c2})$ increases the inefficiency of parallel flow decreases.

Example 9.3
A shell and tube heat exchanger is required to raise 0.5 kg/s of LTHW from 70°C to 85°C.

a) Determine the heating surface required if 0.0158 kg/s of steam at 4 bar absolute and 0.9 dry is used as the primary medium in parallel flow.
Data: overall heat transfer coefficient $U = 1.2\,\text{kW/m}^2\text{K}$, specific heat capacity for water $C = 4.18\,\text{kJ/kgK}$.
b) What effect does counterflow have upon the surface area?

Solution (a)
Use will be made of the tables of *Thermodynamic and Transport Properties of Fluids* in the solution.

The output of the heat exchanger can be obtained from the secondary side and

$$Q = M_c C_c dt = 0.5 \times 4.18 \times (85 - 70) = 31.35\,\text{kW}.$$

Also $Q = M_h(h_w - h_f)\,\text{kW}$

and from the tables $h_w = (h_f + q h_{fg}) = 605 + 0.9 \times 2134 = 2526\,\text{kJ/kg}$.

Substituting: $31.35 = 0.015\,83(2526 - h_f)$

from which $h_f = 546\,\text{kJ/kg}$

from the tables therefore $t_s = t_{h2} = 130°\text{C}$.

For parallel flow: primary fluid (steam) $143.6 \to 130$
secondary fluid (water) $70 \to 85$

$dt_{\text{max,min}}$ 73.6 45

LMTD $dt_m = (73.6 - 45)/\ln(73.6/45) = 58\,\text{K}$

since $Q = UA dt_m\,\text{kW}$

then heat exchange surface $A = Q/U dt_m = 31.35/1.2 \times 58 = 0.45\,\text{m}^2$.

Solution (b)
Considering counterflow: primary fluid (steam) $143.6 \to 130$
secondary fluid (water) $85 \leftarrow 70$

$t_{\text{min,max}}$ 58.6 60

Since these mean temperatures are closely similar the arithmetic mean can be taken and:

for the primary fluid $t_m = (143.6 + 130)/2 = 136.8°\text{C}$
for the secondary fluid $t_m = (85 + 70)/2 = 77.5°\text{C}$
from which $dt_m = 136.8 - 77.5 = 59.3\,\text{K}$.

You should now determine the LMTD to confirm that it agrees with this arithmetic mean. This closely corresponds to the LMTD of 58 K for parallel flow and therefore will have little influence over the heat exchanger surface.

Example 9.4

Tetrafluoroethane (refrigerant 134a) leaves a compressor at 7.7 bar absolute with 20 K of superheat and enters a condenser at the rate of 0.025 kg/s. The coolant temperature at entry is 12°C at a mass flow rate of 0.08 kg/s. Assuming counterflow determine the heat exchange surface and the output of the condenser.

Data: fouling factor 0.0002 m²K/W,
specific heat capacity of the coolant 4.18 kJ/kgK,
heat transfer coefficient at the inside surface 850 W/m²K,
heat transfer coefficient at the outside surface 600 W/m²K

Solution

Use will be made of the tables of *Thermodynamic and Transport Properties of Fluids* for data relating to the refrigerant.

Considering the primary fluid $Q = M_h(h - h_f)$ W.

From the tables the following data is obtained for refrigerant 134a:

$h = 435.44$ kJ/kg, $h_f = 241.69$ kJ/kg

thus at 7.7 bar absolute $Q = 0.025(435.44 - 241.69) = 4.844$ kW.

Now considering the secondary fluid $Q = M_c C_c dt$.

Substituting: $4.844 = 0.08 \times 4.18(t_{c2} - 12)$

from which $t_{c2} = 26.5°C$.

From the tables the following data is obtained for refrigerant 134a at 7.7 bar absolute:

the superheat temperature is 50°C and the saturation temperature is 30°C:

for counterflow: primary fluid   50 → 30
                secondary fluid   26.5 ← 12

                $dt_{max,min}$ 23.5   18

LMTD $dt_m = (23.5 - 18)/\ln(23.5/18) = 20.63$ K.

The overall heat transfer coefficient $U = 1/R_t$ kW/m²K
and $R_t = (1/h_{si}) + R_f + (1/h_{so}) = (1/0.85) + 0.0002 + (1/0.6) = 2.8433$ m²K/kW

and therefore $U = 1/2.8433 = 0.3517$ kW/m²K.

Given $Q = UA dt_m$ kW

heating surface

$A = Q/U dt_m = 4.844/0.3517 \times 20.63 = 0.668$ m².

The output of the condenser is calculated as 4.844 kW.

9.4 Heat exchanger performance

The performance of heat exchangers with a solid boundary between the primary and secondary fluids depends upon the overall heat transfer coefficient U, which acts as the interface between the two fluids. This interface consists of three elements plus the fouling resistance:

the hot side laminar sublayer
the solid interface or boundary
the cold side laminar sublayer.

The main sources of thermal resistance are the two laminar sublayers. In streamline or laminar flow the laminar sublayers offer appreciable thermal resistance because they have significant thickness through which heat must be conducted. Turbulent flow reduces this thickness and baffles are sometimes employed to increase turbulence. Turbulence can be induced in this way in a fluid when $Re > 2000$. The overall heat transfer coefficient is therefore dependent upon fluid velocity on both sides of the solid boundary and upon the fouling resistance. The laminar sublayer in turbulent flow is considered in Chapter 6.

The outside surface of the heat exchanger may be exposed to a fluid which has a lower specific heat capacity than that of the primary fluid. An example would be in the case of an air heater battery supplied from a low temperature hot water heating system. The primary fluid is water having a specific heat capacity of about 4.2 kJ/kgK compared to that of air which has a specific heat capacity of around 1.0 kJ/kgK.

In order to increase the heat transfer potential, extended finning is adopted on the air side of the battery. This increases the surface area to compensate for the lower specific heat capacity of air. The surface area of the heat exchanger should be a maximum within the limits of cost and size.

Consider a calandria consisting of:

(i) 20 tubes with an inside diameter of 40 mm,
(ii) 80 tubes with an inside diameter of 20 mm.

Both tube bundles will fit into the same size shell of 2.5 m in length. The respective surface areas are:

(i) $A = 20\pi \times 0.04 \times 2.5 = 6.283 \, \text{m}^2$
(ii) $A = 80\pi \times 0.02 \times 2.5 = 12.567 \, \text{m}^2$.

Clearly (ii) surface is to be preferred. The true temperature difference between fluids also has a direct influence upon the output of the calandria since it is the motive force in heat transfer. The minimum true temperature difference should not be less than 20 K for good heat exchange.

There are a number of terms used in relation to heat exchangers which describe their performance and allow comparisons to be made. They include:

Capacity ratio – CR
Effectiveness – E
Number of transfer units – NTU.

Capacity ratio is the ratio of the products of mass flow and specific heat of each of the primary and secondary fluids. The product MC is the thermal capacity of the moving fluid, the units of the terms being: (kg/s) × (kJ/kgK) = kJ/sK = kW/K.

Capacity ratio is the ratio of the smaller product to that of the larger and therefore CR < 1.0.

If $(M_h C_h) > (M_c C_c)$ $CR = (M_c C_c)/(M_h C_h)$,
if $(M_c C_c) > (M_h C_h)$ $CR = (M_h C_h)/(M_c C_c)$,

where CR, being a ratio, is dimensionless.

There are two special cases to consider:

(i) The capacity ratio becomes zero for both boiling and condensing where the units for C are kJ/kgK and if the evaporation or condensation of the fluid occurs at constant temperature the temperature drop is 0 K.
(ii) The capacity ratio becomes unity (1.0) for equal thermal capacities of the primary and secondary fluids.

Effectiveness is the ratio of energy actually transferred to the maximum theoretically possible. Again it depends upon the product of mass flow and specific heat capacity of the primary and secondary fluids in kW/K.

If $M_h C_h > M_c C_c$ $E = (t_{c2} - t_{c1})/(t_{h1} - t_{c1})$,
if $M_c C_c > M_h C_h$ $E = (t_{h1} - t_{h2})/(t_{h1} - t_{c1})$

where E, being a ratio, is dimensionless.

In parallel flow t_{c2} approaches t_{h2} but can never exceed it whereas in counterflow t_{c2} can exceed t_{h2} and hence heat exchange in counterflow can be more effective. Refer to Figures 9.1 and 9.2.

Number of transfer units is the ratio of the product of the overall heat transfer coefficient and heat exchange area, and the thermal capacity MC of either the primary or secondary fluid. The units of the product of the terms UA are (kW/m²K) × m² = kW/K and since these are the same as the product of the terms MC, the number of transfer units, like CR and E, is a dimensionless quantity. The ratio NTU was developed by W. M. Kays and A.L. London and published in 1964.

If $M_h C_h > M_c C_c$ $NTU = UA/M_c C_c$
if $M_c C_c > M_h C_h$ $NTU = UA/M_h C_h$

Capacity ratio, effectiveness and number of transfer units provide a straightforward route in the determination of the leaving temperatures of the primary and secondary fluids t_{h2} and t_{c2}, and in the heat exchanger output Q.

Heat exchanger performance

For counterflow heat exchangers:

$$E = [1 - \exp(-NTU(1 - CR))]/[1 - CR\exp(-NTU(1 - CR))]$$

when $CR = 0$, $E = [1 - \exp(-NTU)]$
when $CR = 1$, $E = NTU/(NTU + 1)$.

For parallel heat exchangers:

$$E = [1 - \exp(-NTU(1 + CR))]/(1 + CR)$$

when $CR = 0$, $E = [1 - \exp(-NTU)]$
when $CR = 1$, $E = [1 - \exp(-2NTU)]/2$.

There now follows some examples using the heat exchanger indices described above.

Example 9.5
(a) Determine the effectiveness and fluid outlet temperatures of an economizer handling 0.8 kg/s of flue gas at an inlet temperature of 280 °C. The mean specific heat capacity is 1.02 kJ/kgK and boiler feed water entering at 0.6 kg/s and 60 °C passes in parallel flow. The heat transfer surface is 1.8 m² and the overall heat transfer coefficient is known to be 1.85 kW/m²K. Take the mean specific heat capacity of the feed water as 4.24 kJ/kgK.
(b) What likely effect would a counterflow heat exchanger have on the flue gas?

Solution (a)
Figure 9.5 shows a typical arrangement in which the economizer is used to extract heat from the boiler flue gases for heating the boiler feed water. By determining the NTU and the CR for the economizer, the effectiveness (E) of its parallel flow heat exchanger can be evaluated. First of all, however, the products of mass flow and specific heat capacity of the primary and secondary fluids must be calculated.

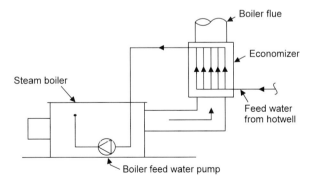

Figure 9.5 Location of the economizer for Example 9.5.

198 Regimes of fluid flow in heat exchangers

$$M_h C_h = 0.8 \times 1.02 = 0.816 \, \text{kW/K}$$
$$M_c C_c = 0.6 \times 4.24 = 2.544 \, \text{kW/K}$$

therefore since the capacity ratio is the smaller thermal capacity over the greater

$$CR = 0.816/2.544 = 0.32$$

and $\quad$ NTU $= 1.85 \times 1.8/0.816 = 4.081$.

Effectiveness E can now be evaluated and
$$E = [1 - \exp(-4.081(1 + 0.32))]/(1 + 0.32)$$
from which $E = (1 - 0.00458)/1.32 = 0.754$.

Note: $\exp(-4.081(1 + 0.32)) = 2.7183^{-5.3869} = 0.00458$.

Since $E = (t_{h1} - t_{h2})/(t_{h1} - t_{c1})$
then $0.754 = (280 - t_{h2})/(280 - 60)$
from which $t_{h2} = 114°C$.

Using the heat balance: heat lost by flue gas = heat gain by feed water
$$0.816(280 - 114) = 2.544 \, dt$$
from which the temperature rise in the feed water $dt = 53 \, K$
and therefore $t_{c2} = 53 + 60 = 113°C$.

Solution (b)
For counterflow
$$E = [1 - \exp(-4.081(1 - 0.32))]/[1 - 0.32\exp(-4.081(1 - 0.32))]$$
from which $\quad E = (1 - 0.0623)/(1 - 0.01995)$
and $\quad\quad\quad E = 0.957$.

This shows that counterflow is more effective than parallel flow where $E = 0.754$.

Since $M_c C_c > M_h C_h \quad E = (280 - t_{h2})/(280 - 60)$
from which $\quad t_{h2} = 70°C$.

Adopting the heat balance: heat lost by flue gas = heat gain by feed water
$$0.816(280 - 70) = 2.544 \, dt$$
from which the temperature rise in the boiler feed water $dt = 67 \, K$, and the leaving temperature $t_{c2} = 67 + 60 = 127°C$.

Summary for Example 9.5
Clearly the counterflow heat exchanger is more effective. However, too much heat may be being extracted from the flue gas as t_{h2} at 70°C is likely to be below the dew point of the flue gas and corrosion would have to be accounted for in the chimney.

Example 9.6

0.18 kg/s of steam at 3.5 bar absolute, 0.9 dry enters a counterflow heat exchanger serving an HWS storage calorifier and condensate leaves at a temperature of 138.9°C. Feed water enters the calorifier at 10°C at the rate of 1.5 kg/s to satisfy the simultaneous HWS demand.

The heat transfer coefficients at the inside and outside surfaces of the heat exchanger are 13 kW/m²K and 10 kW/m²K respectively. Fouling resistance and the thermal resistance of the solid boundary between the primary and secondary fluids may be ignored.

Given the specific heat capacity of water as 4.2 kJ/kgK, determine, for the heat exchanger in the HWS calorifier, its capacity ratio, effectiveness, number of transfer units and heat exchange surface.

Solution
The tables of *Thermodynamic and Transport Properties of Fluids* will be needed for the solution. From the tables it can be seen that there is no change in temperature of the primary steam which gives up its latent heat only in the heat exchanger. The larger thermal capacity (where for C here in kJ/kgK, temperature difference K being zero) is infinite. Therefore the capacity ratio CR = 0.

Using the heat balance to evaluate the leaving temperature of the secondary water t_{c2} heat lost by the primary steam = heat gain by the secondary water

$$M_h(qh_{fg}) = M_c C_c dt.$$

Note from the tables that the heat lost by the steam takes place at a constant temperature of 138.9°C. This therefore is a case of condensation at constant temperature. Refer to Figure 9.4.

Substituting from the tables and the data in the question:

$$0.18 \times (0.9 \times 2148) = 1.5 \times 4.2(t_{c2} - 10)$$

from which the secondary outlet temperature $t_{c2} = 65.23°C$. Since the primary fluid is at constant temperature the arithmetic mean temperature difference between the fluids is the true mean value and

$$dt_m = 138.9 - (65.23 + 10)/2 = 101.29 \, K$$

The overall heat transfer coefficient $U = 1/R_t$ kW

where $R_t = (1/h_{si}) + (1/h_{so}) = (1/13) + 1/10 = 0.1769 \, m^2K/W$
and $U = 1/0.1769 = 5.652 \, kW/m^2K$.

Now the rate of heat transfer at the heat exchanger can be obtained from either side of the heat balance,

thus $Q = 0.18 \times (0.9 \times 2148) = 348 \, kW$.
Since $Q = UAdt$

heat exchanger surface
$A = Q/U dt = 348/5.652 \times 101.29 = 0.608 \text{ m}^2$.

NTU $= UA/M_c C_c$ since the secondary water has the lower thermal capacity as the thermal capacity of the primary steam is zero,

so NTU $= 5.652 \times 0.608/1.5 \times 4.2 = 0.545$.

Since capacity ratio CR $= 0$, the effectiveness of the heat exchanger

$E = 1 - \exp(-\text{NTU}) = 1 - 2.7183^{(-0.545)} = 0.42$.

Summary for Example 9.6
Capacity ratio CR $= 0$, effectiveness $E = 0.42$, number of transfer units NTU $= 0.545$ and the heat exchange surface $A = 0.608 \text{ m}^2$.

9.5 Cross flow

Figure 9.6 shows a typical air heater battery through which the primary fluid is constrained within the heat transfer tubes and over which

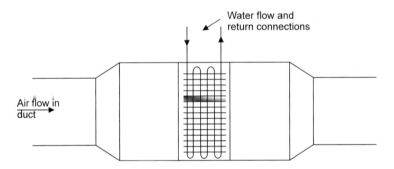

Figure 9.6 A cross flow air heater battery.

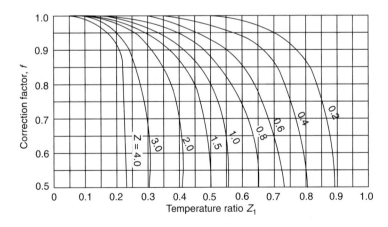

Figure 9.7 LMTD correction factor for cross flow, one fluid mixed, one fluid unmixed.

air flows freely. The primary fluid is unmixed as it is contained within the tube boundary walls while the air which is the secondary fluid is considered as mixed flow. This requires the introduction of a correction factor f in the equation for the heat transfer across the heat exchanger. Thus $Q = UAf dt_m$ kW.

The LMTD dt_m is calculated for cross flow in the same way as for counterflow. Figure 9.7 shows how correction f can be evaluated for a cross flow heat exchanger with one fluid mixed and the other unmixed. When applying the factor it does not matter whether the hotter fluid is mixed or unmixed.

Example 9.7
An air heater battery is supplied with water at 1.0 kg/s and 82°C with return water at 72°C. Air enters the heater at 1.86 m³/s and 20°C.

(a) Assuming cross flow determine correction factor f from Figure 9.7 and hence calculate the heat exchanger surface.
(b) Determine the capacity ratio, effectiveness and the number of transfer units for the heater battery.

Data: specific heat capacities for water and air are 4.2 and 1.0 kJ/kgK respectively, the heat transfer coefficients at the inside and outside surfaces of the exchanger tubes are 3.72 and 2.0 kW/m²K, fouling resistance is 0.0002 m²K/W.

Solution (a)
The horizontal axis of Figure 9.7, $Z_1 = (t_{h2} - t_{h1})/(t_{c1} - t_{h1}) = (70 - 82)/(20 - 82)$, thus $Z_1 = (-12)/(-62)$, from which the temperature ratio $Z_1 = 0.19$.

From the same figure $Z_2 = (t_{c1} - t_{c2})/(t_{h2} - t_{h1}) = (20 - 50)/(70 - 82) = (-30)/(-12)$ from which $Z_2 = 2.5$.

Adopting the values for temperature ratios Z_1 and Z_2, from Figure 9.7 the correction factor $f = 0.96$.

The overall heat transfer coefficient $U = 1/R_t$ kW, and $R_t = (1/3.72) + 0.0002 + (1/2) = 0.769$ m²K/kW, then $U = 1/0.769 = 1.3$ kW/m²K.

Taking cross flow as counterflow to obtain LMTD without loss of integrity:

primary fluid: water 82⟶70
secondary fluid: air 50⟵20
$dt_{max,min}$ 32   50

LMTD $dt_m = (50 - 32)/\ln(50/32) = 40.3\,\text{K}$
From $Q = UAdt_mf$, $A = Q/Udt_mf$
where $Q = M_h C_h dt = 1.0 \times 4.2 \times (82 - 70) = 50.4\,\text{kW}$

and substituting:

heating surface $A = 50.4/1.3 \times 40.3 \times 0.96 = 1.0021\,\text{m}^2$.

Solution (b)

Now $M_h C_h = 1.0 \times 4.2 = 4.2\,\text{kW/K}$
and $M_c C_c = (1.86 \times 0.9) \times 1.0 = 1.674\,\text{kW/k}$.
The capacity ratio $CR = 1.674/4.2 = 0.399$
Number of transfer units $NTU = UA/M_c C_c = 1.3 \times 1.0021/1.674 = 0.778$.

The determination of effectiveness for a cross flow heat exchanger, one fluid mixed, one fluid unmixed, is derived from the relationship between capacity ratio and number of transfer units. An approximate value for effectiveness here may be calculated assuming counterflow. Approximate effectiveness, taking cross flow as counterflow,

$E = [1 - \exp(-0.778(1 - 0.399))]/[1 - 0.399$
$\exp(-0.788(1 - 0.399))] = (1 - 0.6198)/(1 - 0.2473) = 0.505$.

The actual answer is $E = 0.484$.

Summarizing Example 9.7
(a) heating surface $A = 1.0021\,\text{m}^2$.
(b) $CR = 0.399$, $E = 0.505$, $NTU = 0.778$.

The solution must be qualified to the extent that the true temperature difference between fluids and the effectiveness was determined for counterflow.

9.6 Further examples

Example 9.8
A shell and tube non-storage heating calorifier operates in counterflow. The primary medium is high temperature hot water at temperatures of 160°C and 130°C. The secondary medium is low temperature hot water operating at 82°C and 70°C. The heat exchange tube bundle consists of four copper tubes each 20 mm inside diameter, 25 mm outside diameter having a thermal

conductivity of 0.35 kW/mK and heat transfer coefficients of 5 and 3 kW/m²K respectively. Ignoring the effects of fouling on the heat exchange surfaces, determine the mass flow of the secondary medium and the length of the tube bundle. Take the specific heat capacities of the primary and secondary mediums at the appropriate mean water temperature and the mass flow of high temperature hot water as 0.349 kg/s.

Solution
The shell and tube heat exchanger is shown in Figure 9.8.

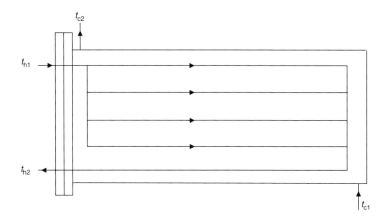

Figure 9.8 Non-storage calorifier (Example 9.8).

The mean water temperature of the primary medium $= (160 + 130)/2 = 145°C$.

The mean water temperature of the secondary medium $= (82 + 70)/2 = 76°C$.

From the tables of *Thermodynamic and Transport Properties of Fluids* the specific heat capacities at these mean water temperatures are:

primary medium, 4.3 kJ/kgK and secondary medium, 4.194 kJ/kgK.

Adopting the heat balance:

heat lost by the primary medium = heat gain by the secondary medium

substituting:

$$0.349 \times 4.3 \times (160 - 130) = M \times 4.194 \times (82 - 70)$$

from which the mass flow of low temperature hot water is 0.895 kg/s.

For counterflow, primary fluid: 160⟶130
secondary fluid: 82⟵ 70
$dt_{max,min}$ 78   60

LMTD $dt_m = (78 - 60)/\ln(78/60) = 68.6\,\text{K}$.

Since the surface of the heat exchanger is identified as four copper tubes it is convenient to determine the overall heat transfer coefficient for radial conductive heat flow which is introduced in Chapter 2.

Thus from Chapter 2: $Q/L = (2\pi dt)/[(1/r_1 h_{si}) + ((\ln r_2/r_1)/k_1) + (1/r_2 h_{so})]\,\text{W/m}$

since $Q/L = U_L dt\,\text{W/m}$ therefore $U_L = (Q/L)/dt\,\text{W/mK}$

then $U_L = (2\pi)/[(1/r_1 h_{si}) + ((\ln(r_2/r_1))/k_1) + (1/r_2 h_{so})]\,\text{W/mK}$.

Note the different units for the overall heat transfer coefficient which is in per metre run of pipe and not per square metre of plane surface.

substituting: $U_L = (2\pi)/[(1/0.01 \times 5) + (\ln(0.0125/0.01)/0.35) + (1/0.0125 \times 3)]$

from which $U_L = (2\pi)/[20 + 0.6376 + 26.667]$

thus $U_L = 0.133\,\text{kW/mK}$.

The output of the heat exchanger can be obtained from either side of the heat balance:

thus $Q = 0.349 \times 4.3 \times (160 - 130) = 45\,\text{kW}$
now $Q/L = U dt\,\text{W/m}$ and therefore $Q = U dt L\,\text{W}$,

therefore

tube bundle length $L = (Q/U_L dt_m) \times 1/4 = (45/0.133 \times 68.6) \times 1/4 = 1.233\,\text{m}$.

Summary for Example 9.8
The mass flow of low temperature hot water is 0.85 kg/s and the length of the tube bundle is 1.233 m which does not account for the inefficiency of heat exchange. Note the low thermal resistance of the copper tubes relative to the heat transfer coefficients at the inner and outer surfaces. This is the reason why the thermal resistance of the solid boundary is often ignored. If mild steel heat exchange tubes were used the thermal conductivity of the material would be in the region of 0.05 kW/mK and the thermal resistance of the solid boundary would be:

$R = (\ln(0.0125/0.01)/0.05) = 4.463\,\text{m}^2\text{K/W}$. This compares with that for copper tubes of $R = 0.6376\,\text{m}^2\text{K/W}$. You can see why copper is favoured as the heat exchange material. The effect on the length of the tube bundle if steel tube is employed for the heat exchanger is $L = 1.333\,\text{m}$ compared with $L = 1.233\,\text{m}$ for copper. You should now confirm the length of the tube bundle for steel.

Example 9.9

Tetrafluoroethane is discharged from a compressor at 14.91 bar absolute having 20 K of superheat and enters a condenser at 0.3 kg/s. It leaves the condenser sub-cooled by 5 K. The coolant flow rate is 0.794 kg/s and the inlet temperature is 16°C. Assuming counterflow determine the condenser output, the leaving temperature of the coolant and the length of the tube bundle.

Data: the tube bundle consists of 8 × 15 mm nominal bore tubes having an outer diameter of 20 mm, the overall heat transfer coefficient is 3.2 kW/m²K and the specific heat capacity of the coolant is 2.8 kJ/kgK.

Solution
Reference should be made to the tables of *Thermodynamic and Transport Properties of Fluids*. From the tables Refrigerant 134a at 14.91 bar absolute has a specific enthalpy of 449.45 kJ/kgK at 20° of superheat. The temperature of the superheated vapour is 75°C and at saturated conditions it is 55°C. Thus since it is sub-cooled by 5 K on leaving the condenser the leaving temperature of the refrigerant will be $55 - 5 = 50°C$. The specific enthalpy of the refrigerant leaving the condenser will therefore be 271.61 kJ/kgK. You should now confirm these data from the tables.

The condenser output $Q = M(h - h_f)$ kW

substituting: $Q = 0.3(449.45 - 271.61) = 53.35$ kW.

Adopting the heat balance:

heat lost by refrigerant = heat gain by coolant

substituting: $53.35 = 0.794 \times 2.8 dt$

from which the temperature rise of the coolant $dt = 24$ K

therefore the leaving temperature of the coolant $t_{c2} = 16 + 24 = 40°C$.

For counterflow: primary fluid 75⟶50
secondary fluid 40⟵16
$dt_{max,min}$ 35   34

LMTD $dt_m = (35 - 34)/\ln(35/34) = 34.5$ K.

Note: since the maximum and minimum temperature differences are almost equal the arithmetic mean temperature difference can be taken and equals $(62.5 - 28) = 34.5$ K.

For the determination of tube bundle length $Q = U_L L dt_m$ kW where U_L has the units kW/m.runK.

Now the overall heat transfer coefficient $U = 3.2$ kW/m²K.

One metre of 20 mm outside diameter tube has an area of:

$$a = \pi D \times 1.0 = \pi \times 0.02 \times 1.0 = 0.0628 \, \text{m}^2$$

therefore $U_L = 0.0628 \times U = 0.0628 \times 3.2 = 0.2 \, \text{kW/mK}$

thus from $Q = U_L L dt_m \, \text{kW}$, $L = Q/U_L dt_m \, \text{m}$.

If there are eight tubes in the bundle

$$L = (Q/U_L dt_m) \times 1/8 = (53.35/0.2 \times 34.5) \times 1/8 = 0.966 \, \text{m}.$$

Summary for Example 9.9
Condenser output 53.35 kW, leaving temperature of the coolant 40°C, length of the tube bundle 0.966 m.

The heat exchanger will be similar to Figure 9.8 but would have eight tubes in the bundle instead of the four shown in the diagram.

9.7 Chapter closure

This completes the work on heat exchangers only a few of which have been considered in detail here. However, the principles of heat exchanger design and performance have been introduced from which you will appreciate that the subject is very specialized and largely in the domain of the manufacturer. It is important though for the student in building services to have some knowledge of this work and you should now extend it by undertaking market research into the types of heat exchanger available.

Verifying the form of an equation by dimensional analysis 10

Nomenclature

A	area (m²)
C	specific heat capacity (kJ/kgK)
d	diameter (m)
dt	temperature difference (K)
f	frictional coefficient
g	gravitational acceleration (m/s²)
h	head (m)
I	flux density (W/m²)
L	length (m), dimension for length (m)
LHS	left-hand side
M	mass flow (kg/s), dimension for mass (kg)
P	pressure (Pa)
Q	rate of heat flow (W), volume flow (m³/s)
RHS	right-hand side
T	dimension for time (s)
TE	total energy, metres of fluid flowing
U	thermal transmittance coefficient (W/m²K)
u	velocity (m/s)
θ	dimension for thermodynamic temperature (K)
μ	absolute viscosity (kg/ms)
ρ	density (kg/m³)
σ	Stefan–Boltzmann constant (W/m²K⁴)

10.1 Introduction

The process of checking the units of the terms in an equation is relatively straightforward and a common strategy for ensuring that its form is correct. For example the formula $Q = UA\mathrm{d}t$ has the units of watts. This unit for the term Q can be checked if the units of the other terms are known, thus the product of the terms $UA\mathrm{d}t$ has the units (W/m²K) × (m²) × (K) and by the process of cancellation the unit of watts is confirmed.

Dimensions on the other hand are properties (of a term) which can be measured. For example, density has the units kg/m^3. Its dimensions are ML^{-3} where M is mass and L is length.

Units are the elements by which numerical values of these dimensions describe the term quantitatively. That is to say the units of density are kg/m^3 where mass is quantified in kg and volume is quantified in m^3. Thus the units of a term define in addition the system of units being used. A term's dimensions on the other hand are not confined to any system of units and therefore dimensional analysis is universal and common to all systems of units.

There are three systems of units in use in the western world namely: Systeme International SI, FPS or foot-pound-second system and MKS or metre-kilogram-second system. The UK has for some years adopted the Systeme International. The United States of America currently uses the FPS system and Germany the MKS system of measurement.

Dimensional analysis is adopted to undertake three discrete tasks which are:

- to check that an equation has been correctly formed
- to establish the form of an equation relating a number of variables
- to assist in the analysis of empirical formulae in experimental work.

This chapter will focus on checking some equations used in this book by dimensional analysis to show that they are correctly formed.

10.2 Dimensions in use

It is necessary to identify the dimensions which will be used in validating a formula. There are four dimensions which are basic to heat and mass transfer and they are those for mass M, length L, time T and thermodynamic temperature θ.

The dimensions for force, for example, may now be established and since Force = mass × acceleration, the units of the terms are: force = kg × m/s^2. The dimensions of the terms are: force = $M \times L \times T^{-2} = MLT^{-2}$. The unit of force in the Systeme International is the newton. thus the dimensions of the newton are MLT^{-2}.

The newton is called a derived unit since it is made up from more than one basic unit. There are six basic units in the Systeme International and the dimensions of four of them, M, L, T and θ, will be used to define the dimensions of terms in this chapter. You can see the importance of knowing the basic units in a derived unit such as the newton before identifying its dimensions. There are a number of derived units of terms commonly used in heat and mass transfer. They are listed in Table 10.1 and you should confirm the units and dimensions of each one.

There now follow some examples in which the forms of some equations used in this book are checked by dimensional analysis.

Table 10.1 Dimensions of some derived units used in heat and mass transfer

Term	Name	Symbol	Units, definitions	Dimensions
Force	newton	F	kg × m/s²	MLT^{-2}
Energy	joule (Nm)	H	force × distance moved	ML^2T^{-2}
Power	watt	P_w, Q	energy/time	ML^2T^{-3}
Pressure	pascal	P	force/area	$ML^{-1}T^{-2}$
Gravitational acceleration		g	m/s²	LT^{-2}
Density		ρ	kg/m³	ML^{-3}
Mass flow		M	kg/s	MT^{-1}
Volume flow		Q	m³/s	L^3T^{-1}
Absolute viscosity		μ	kg/ms	$ML^{-1}T^{-1}$
Kinematic viscosity		v	absolute viscosity/density	L^2T
Specific heat capacity		C	J/kgK	$L^2T^{-2}\theta^{-1}$
Specific enthalpy		h	kJ/kg	L^2T^{-2}
Cubical expansion		β	1/θ	θ^{-1}
Thermal conductivity		k	W/mK	$MLT^{-3}\theta^{-1}$
Heat flux		I	W/m²	MT^{-3}
Speed of rotation		N	rev/s	T^{-1}
Heat transfer coefficient		h	W/m²K	$MT^{-3}\theta^{-1}$
Mean velocity		u	m/s	LT^{-1}
Area		A	m²	L^2

Example 10.1
$P = h\rho g$ Pa.

Solution
Refer to Table 10.1.
The right-hand side of the equation has the dimensions $(L)(ML^{-3})(LT^{-2})$, these reduce to $ML^{-1}T^{-2}$ which agrees with the dimensions for pressure in Table 10.1.

Example 10.2
$h = (4fLu^2)/(2gd)$ m of fluid flowing.

Solution
Refer to Table 10.1.
The right-hand side of the formula has the dimensions $[(L)(L^2/T^2)/(L/T^2)(L)]$. Note that the pure numbers do not have dimensions. It is assumed for the moment that the coefficient of friction f is dimensionless.
The dimensions on the RHS of the formula reduce to L. This shows that the form of the Darcy formula is correct and confirms that the coefficient of friction is dimensionless.

Example 10.3
Show by dimensional analysis that the Reynolds number is dimensionless.

Solution
Now $Re = (\rho u d)/\mu$.
Referring to Table 10.1 and substituting the dimensions on the RHS of the equation: $(ML^{-3})(LT^{-1})(L)/(ML^{-1}T^{-1}) = (ML^{-3})(LT^{-1})(L)(M^{-1}LT)$ from which all the dimensions on the RHS cancel and therefore the Reynolds number is dimensionless.

Example 10.4
Check the form of Box's formula for head loss in turbulent flow in straight pipes where: $h = (fLQ^2)/(3d^5)$ m of fluid flowing.

Solution
Referring to Table 10.1 and substituting the dimensions on the RHS of the formula:

$$(L)(L^3T^{-1})^2(L^{-5}) = (L)(L^6T^{-2})(L^{-5})$$

the RHS reduces to: $(L^2)(T^{-2})$.

The RHS should reduce to dimension (L) to equate with the LHS of the formula. This leaves the dimensions $(L)(T^{-2})$ and from Table 10.1 the term which will cancel these dimensions is the inverse of gravitational acceleration which is $(T^2)(L^{-1})$. The conclusion therefore is that the constant (1/3) in the formula must include the term gravitational acceleration in its denominator.

From Chapter 7, in the derivation of Box's formula,

$$h = (64/2\pi^2 g)(fLQ^2/d^5) \text{ metres of fluid flowing,}$$

and gravitational acceleration g appears in the denominator of the constant which is enclosed by the first set of brackets. If g is taken as 9.81 m/s^2 the constant evaluates to 1/3. The analysis of dimensions therefore has shown that the form of Box's formula is correct. It also identifies the fact that the constant in the formula of (1/3) does have dimensions which are $(T^2)(L^{-1})$.

Example 10.5
Show that the equation for the determination of mass flow of water $Q = MCdt$, is formed correctly.

Solution
From Table 10.1 the dimensions of Q which is the rate of heat flow in watts are: ML^2T^{-3}.

The dimensions of the terms on the RHS of the equation are, where specific heat capacity from Table 10.1 has the dimensions $(L^2T^{-2}\theta^{-1})$: $(MT^{-1})(L^2T^{-2}\theta^{-1})(\theta) = (ML^2T^{-3})$ which agrees with the LHS of the equation and therefore the form of the equation is correct.

Example 10.6
Show that the formula $TE = Z + (P\rho g) + (u^2/2g)$ metres of fluid flowing is formed correctly.

Solution
Refer to Table 10.1. Total energy

$$TE = (L) + (ML^{-1}T^{-2})/(ML^{-3})(LT^{-2}) + (L^2T^{-2})/LT^{-2}),$$

from which $TE = (L) + (L) + (L)$ and the formula for total energy is shown to be formed correctly.

Example 10.7
Adopting dimensional analysis verify the dimensions of the Stefan–Boltzmann constant for heat radiation.

Solution
The equation is flux density $I = \sigma T^4$ W/m^2.

Referring to Table 10.1, the dimensions of heat flux on the LHS of the equation are: MT^{-3}.

The RHS of the formula is the product of a numerical constant and thermodynamic temperature to the fourth power thus: σT^4.

In order for the right-hand side of the formula to be dimensionally similar to the left-hand side the product of the Stefan–Boltzmann constant and thermodynamic temperature to the fourth power must have the dimensions: $(MT^{-3}\theta^{-4})(\theta^4)$. This will then reduce the RHS of the formula to MT^{-3} which is dimensionally similar to the LHS. Thus the dimensions of the Stefan–Boltzmann constant will be: $MT^{-3}\theta^{-4}$.

You will note that not all numerical constants in formulae are dimensionless. Example 10.4 is another case in point.

10.3 Chapter closure

You have now been introduced to the dimensions of terms in common use and undertaken one of the applications to which dimensional analysis can be put. Clearly it is easier to check the form of an equation by using the units of the terms within it as shown in section 10.1, rather than by using the dimensions of the terms. However it is important to persevere with using the dimensions of terms since they are used exclusively in Chapter 11.

This chapter has sought to illustrate the way the dimensions of terms are expressed and provides a grounding for the next chapter which is where dimensional analysis has a major role in establishing the form of an equation relating a number of variables and in assisting the analysis of experimental work.

Solving problems by dimensional analysis 11

Nomenclature

Some of the nomenclature used in this chapter is given in Table 10.1 of Chapter 10 to which reference will need to be made. The remainder is listed here.

- B constant of proportionality
- d diameter, distance from leading edge (m)
- dh difference in head (m of fluid flowing)
- dP pressure drop (Pa)
- f frictional coefficient for turbulent flow
- Gr Grashof number
- L length (m)
- m number of dimensions
- n number of variables
- Nu Nusselt number
- Pr Prandtl number
- Re Reynolds number
- ϕ function of
- τ shear stress (Pa)

11.1 Introduction

Chapter 10 introduces one of the three applications of dimensional analysis. The two remaining applications include: establishing the form of an equation relating a number of variables and assisting in establishing empirical formulae in experimental work.

The dimensions which will be used in this chapter continue to be those for mass M, length, L, time T and thermodynamic temperature θ. The dimension for heat energy H is used in Example 11.7.

11.2 Establishing the form of an equation

A formula which is well known to building services will be used now to demonstrate this use of dimensional analysis.

Case study 11.1

Consider the Darcy equation for turbulent flow which may be written as:

$dh = (4fLu^2)/2gd$ m of fluid flowing.

This formula may be in terms of units of pressure and since $dP = dh\rho g$ Pa the Darcy equation can be expressed as

$dP = [(4fLu^2)/2gd]\rho g$ Pa

thus $dP = (4fLu^2\rho)/2d$ Pa.

If it is required to express the formula in terms of a rate of pressure loss per metre:

then $dP/L = (4fu^2\rho)/2d$ Pa/m.

We will now use dimensional analysis to find the form of an equation for turbulent flow using a number of variables.

With turbulent flow of a fluid in a flooded pipe, pressure loss per unit length is likely to be related to the variables u, ρ, μ and a characteristic dimension of the pipe or duct, say diameter d,

thus $dP/L = \phi(u, \rho, \mu, d)$.

The term ϕ means 'a function of'.

If an exponential relationship is assumed

$$dP/L = B[u^a \rho^b \mu^c d^d] \qquad (11.1)$$

where B is taken as a constant of proportionality and the indices a, b, c and d may need to be evaluated empirically.

The expression is now changed to the dimensions of the terms within it and using Table 10.1 to assist in defining the terms:

$$(ML^{-2}T^{-2}) = B[(LT^{-1})^a (ML^{-3})^b (ML^{-1}T^{-1})^c (L)^d]$$

Now forming equations from the indices and remembering, for example, that the dimension M is in fact M^1:

for dimension M: $1 = b + c$
for dimension L: $-2 = a - 3b + c + d$
for dimension T: $-2 = -a - c$

from which $b = 1 - c$, $a = 2 - c$ and $d = -1 - c$.

Substituting the expressions for a, c and d into equation (11.1)

$$dP/L = B[u^{(2-c)} \rho^{(1-c)} \mu^c d^{(-1-c)}].$$

Inspecting this statement identifies from the indices that there are two groups of variables on the RHS of the equation namely one with a numerical index and one with the index $-c$. Putting the terms into these two groupings we have:

$$dP/L = B[(u^2 \rho/d)(\rho u d/\mu)^{-c}].$$

You will notice that the last group of terms $(\rho u d/\mu)^{-c}$ is in fact the Reynolds number Re, and it is found by experiment that the constant of proportionality $B =$ unity, thus we have $dP/L = (u^2\rho/d)\phi(Re)$, the term ϕ accounting for the index $-c$.

Now the Darcy equation may be expressed in the form:

$$dP/L = 4fLu^2\rho/2d = (u^2\rho/d)(4f/2)$$

from which therefore $\phi(Re) = 4f/2$.

The frictional coefficient f is obtained from the Moody diagram which has a scale of Reynolds numbers as one of its axes. Refer to Chapter 6, Figure 6.13.

SUMMARY FOR CASE STUDY 11.1

It is evident from the result of the analysis of the dimensions of the selected variables that the equation so derived is comparable with the Darcy formula for turbulent flow.

BUCKINGHAM'S PI THEOREM

This states that if there are n variables in a problem with m dimensions there will be $(n - m)$ dimensionless groups in the solution. Applying this theorem to case study 11.1, from equation (11.1) there are $n = 5$ variables. The term (dP/L) which is the subject of the expression is taken as one variable, the others are u, ρ, μ and d.

Subsequently three dimensions (M, L, T) were used in the analysis. Thus $(n - m) = (5 - 3) = 2$ dimensionless groups.

The Reynolds number has been identified as one of the groups. The other group can now be found thus:

$$dP/L = (u^2\rho/d)\theta(Re)$$
then $\theta(Re) = (dP/L)(d/u^2\rho)$

and the second dimensionless group must be $(dPd/u^2\rho L)$, in order to keep the equation in balance. You should now check these dimensions of the terms to establish that this group is dimensionless.

11.3 Dimensional analysis in experimental work

Where full-scale experiments cannot be conducted because of the problem of size, information can be obtained by experiments on models provided that the model is related properly to the full-size counterpart. The relationship is simple if fluid flow, for example, is geometrically and dynamically similar in each case.

GEOMETRICAL SIMILARITY

This is achieved when one system is the scale model of the other; that is the ratio of corresponding lengths is constant.

Solving problems by dimensional analysis

DYNAMICAL SIMILARITY

This is achieved when several forces acting on corresponding fluid elements have the same ratio to one another in both systems. For example, in turbulent flow in flooded pipes, the value of the Reynolds number must be identical in the scale model and its full-size counterpart. In case study 11.1 both dimensionless groups would need to satisfy this condition if a model of a full-size version of a system involving turbulent flow in horizontal straight pipes was being built,

thus $(Re)_m = (Re)_{fs}$ **and** $(dPd/u^2\rho L)_m = (dPd/u^2\rho L)_{fs}$

where subscript m refers to the model and subscript fs refers to the full-size version.

These dimensionless groups will now be used in the case study which follows.

Case study 11.2

The pressure loss in a pipe 150 mm bore and 33 m long is 20 kPa when water flows at a mean velocity of 2.5 m/s. Determine the pressure loss when sludge flows at the corresponding speed through a pipe 600 mm bore and 400 m long.

Data: density of water and sludge is 1000 kg/m³ and 2000 kg/m³ respectively,
absolute viscosity for water and sludge is 0.001 kg/ms and 0.003 kg/ms respectively.

SOLUTION

The formula appropriate to this solution is the one analysed in case study 11.1,

$(dPd/u^2\rho L) = \phi(Re)$.

For geometric similarity it is assumed that the water system and the sludge system are to scale.

For dynamic similarity: $(Re)_m = (Re)_{fs}$ and $(dPd/u^2\rho L)_m = (dPd/u^2\rho L)_{fs}$.

For equality of Reynolds numbers

substituting: $(1000 \times 2.5 \times 0.15/0.001)_m = (2000 \times u \times 0.6/0.003)_{fs}$

from which the corresponding mean velocity of the sludge $u = 0.9375$ m/s.

For equality of the second dimensionless group

substituting: $(20 \times 0.15/2.5^2 \times 1000 \times 33)_m$
$= (dP \times 0.6/0.9375^2 \times 2000 \times 400)_{fs}$

from which the corresponding pressure drop = 17 kPa.

11.4 Examples in dimensional analysis

Example 11.1
The laws for centrifugal pumps and fans are: $Q \propto N$, $P \propto N^2$ and $P_w \propto N^3$. Verify these laws using dimensional analysis and identify the dimensions of the numerical constant B.

Solution

Let $Q = \phi(N)$ (11.2)

then $Q = B(N^a)$ where B and a are numerical constants

from Table 10.1 the dimensions of the terms are: $(L^3 T^{-1}) = (T^{-1})$

then $(L^3 T^{-1}) = T^{-a}$

for L $3 = 0$

for T $-1 = -a$ therefore $a = 1$

substituting into equation (11.2) $Q = \phi(N) = B(N)$

if B is the constant of proportionality, $Q \propto N$.

Identifying the dimensions of the constant B from $Q = B(N)$, thus $(L^3 T^{-1}) = B(T^{-1})$

from which the dimensions of the numerical constant $B = L^3$.

Let $P = \phi(N)$ (11.3)

then $P = B(N^a)$.

From Table 10.1 the dimensions of the terms are: $(ML^{-1}T^{-2}) = (T^{-1})$

then $(ML^{-1}T^{-2}) = T^{-a}$

for M $1 = 0$

for L $-1 = 0$

for T $-2 = -a$ from which $a = 2$

substituting into equation (11.3): $P = \phi(N^2) = B(N^2)$,

if B is the constant of proportionality, $P \propto N^2$.

Identifying the dimensions of the constant B from $P = B(N^2)$,

thus $(ML^{-1}T^{-2}) = B(T^{-2})$

from which the dimensions of the numerical constant $B = (ML^{-1})$.

Let $P_w = \phi(N)$ (11.4)

then $P_w = B(N^a)$.

From Table 10.1 the dimensions of the terms are:
$(ML^2T^{-3}) = B(T^{-1})$

then $(ML^2T^{-3}) = T^{-a}$

for M $1 = 0$

for L $2 = 0$

for T $-3 = -a$ from which $a = 3$.

Substituting into equation (11.4): $P_w = \phi(N^3) = B(N^3)$,

if B is the constant of proportionality, $P_w \propto N^3$.

Identifying the dimensions of the constant B from $P_w = B(N^3)$

thus $ML^2T^{-3} = B(T^{-3})$

from which the dimensions of the numerical constant $B = (ML^2)$.

Summarizing Example 11.1

Table 11.1 confirms the relationship between the variables in Example 11.1 and identifies the dimensions of B, the numerical constants in the solutions. In order to remove the dimensions of B it is necessary to consider what other variables may contribute to each of the three equations for Q, P and P_w. There are a number of possibilities when considering prime movers such as centrifugal pumps and fans. They include:

impeller diameter d
impeller speed N
fluid density ρ
absolute viscosity of the fluid μ
kinematic viscosity γ.

By analysing the dimensions of each of these variables from Table 10.1 (the dimension for impeller diameter d being L) it is possible to deduce the term or terms whose dimensions will equate with the numerical constant B. These are shown in the third column of Table

Table 11.1 Summary of Example 11.3; the laws for pumps and fans.

Relationship between variables	Dimensions of numerical constant B	Equivalent terms	Relationship when constant B is dimensionless
$Q \propto N$	L^3	d^3	$Q \propto Nd^3$
$P \propto N^2$	ML^{-1}	ρd^2	$P \propto N^2 d^2 \rho$
$P_w \propto N^3$	ML^2	ρd^5	$P_w \propto N^3 d^5 \rho$

11.1. You should confirm that the equivalent terms under column three in Table 11.1 have dimensions which will cancel the dimensions of the numerical constant in column two of the table.

Thus the numerical constant B is now reduced to a dimensionless number in each case and therefore: $Q = (B)Nd^3$, $P = (B)N^2d^2\rho$ and $P_w = (B)N^3d^5\rho$.

Example 11.2
(a) Show by dimensional analysis that the power P_w required by a fan of diameter d rotating at speed N and delivering a volume per unit time Q of a fluid density ρ and viscosity μ is given by: $P_w = (\rho N^3 d^5), \phi[(Nd^3/Q), (\rho Nd^2/\mu)]$.
(b) Identify the dimensionless groups in the formula.

Solution (a)
Following the procedure in case study 11.1

$$P_w = B[(d^a)(N^b)(Q^c)(\rho^d)(\mu^e)]. \tag{11.5}$$

Assuming an exponential relationship and substituting dimensions for the variables:

$$MLT^{-3} = B[(L)^a(T^{-1})^b(L^3T^{-1})^c(ML^{-3})^d(ML^{-1}T^{-1})^e]$$

for M $\quad 1 = d + e$
for L $\quad 2 = a + 3c - 3d - e$
for T $\quad -3 = -b - c - e$

thus $d = 1 - e$
and $a = 2 - 3c + 3(1 - e) + e$
from which $a = 5 - 3c - 2e$
also $b = 3 - c - e$

Substituting the indicial equations for a, b, and d into equation (11.5):

$$P_w = B[(d^{(5-3c-2e)})(N^{(3-c-e)})(Q^c)(\rho^{(1-e)})(\mu^e)]$$

The indices identify three groups: numerical, $-c$ and $-e$. Putting the terms into these groupings, we have:

$$P_w = B[(\rho N^3 d^5), (Nd^3/Q)^{-c}, (\rho Nd^2/\mu)^{-e}].$$

Note that in the last group which has the index $-e$, μ goes in the denominator since its index is $+e$. This is also the case for Q in the second group. Also note that the constant B and the indices $-c$ and $-e$ can be replaced with the 'function of' term ϕ.

Thus $P_w = (\rho N^3 d^5), \phi[(Nd^3/Q), (\rho Nd^2/\mu)]$.

Solution (b)
Applying Buckingham's Pi theorem, there are six variables and three dimensions, thus there are $(n - m) = (6 - 3) = 3$ dimensionless groups. By analysing the dimensions of the second and third group on the right-hand side of the formula these are found to be dimensionless. You should now confirm that this is so. The remaining dimensionless group must therefore be $(P_w/\rho N^3 d^5)$. You should now confirm that this also is the case.

Example 11.3
The performance of a fan is to be estimated using a scale model. The prototype fan duty is $4\,\mathrm{m}^3/\mathrm{s}$ at 8 rev/s and the power absorbed is 1450 W whereas the model absorbs 800 W for a flow of $1.3\,\mathrm{m}^3/\mathrm{s}$. Determine:

(i) the scale of the model
(ii) the corresponding speed of the model.

Assume the density and viscosity of the air are constant.

Solution (i)
Adopting the formula for power P_w required by the fan in Example 11.2 and using the dimensionless groups identified in part (b) of the solution,

$$(d^3 N/Q)_m = (d^3 N/Q)_{fs}$$

thus $N_m = (Q_m/Q_{fs})N_{fs}(d_{fs}^3/d_m^3)$

Substituting: $N_m = (1.3/4) \times 8(d_{fs}^3/d_m^3)$

from which $N_m = 2.6(d_{fs}^3/d_m^3)$.

Also $(P_w/d^5 N^3)_m = (P_w/d^5 N^3)_{fs}$

then $(P_m/P_{fs})N_{fs}^3 = N_m^3(d_m^5/d_{fs}^5)$.

Substituting: $(800/1450) \times 8^3 = N_m^3(d_m^5/d_{fs}^5)$

substituting for N_m $\quad 282.48 = (2.6 d_{fs}^3/d_m^3)^3 (d_m^5/d_{fs}^5)$

from which $\quad 282.48/17.576 = (d_{fs}^9/d_m^9)(d_m^5/d_{fs}^5)$

and therefore $\quad 16.072 = (d_{fs}^4/d_m^4)$

thus $\quad 2 = (d_{fs}/d_m)$

and the prototype is therefore twice the size of the model.

Solution (ii)
From $N_m = 2.6(d_{fs}^3/d_m^3)$

$$N = 2.6(2/1)^3 = 20.8\,\mathrm{rev/s}.$$

Example 11.4

For a fluid flowing in a pipe, the wall shear stress τ is considered dependent upon the following variables: pipe diameter d, fluid density ρ, fluid viscosity μ, the mean fluid velocity u and the mean height of the roughness projections on the pipe wall k_s.

(a) Show by dimensional analysis that $(\tau/\rho u^2) = \phi[(\rho u d/\mu), (k_s/d)]$.
(b) Identify the dimensionless groups in the formula.

Solution (a)
Assuming an exponential relationship where B is the constant of proportionality and the indices a, b, c, d and e are numerical constants:

$$\tau = B[(d^a)(\rho^b)(\mu^c)(u^d)(k_s^e)] \tag{11.6}$$

Shear stress $\tau =$ (force/area) and therefore has the same dimensions as pressure. Refer to Table 10.1 for this and the dimensions of the other terms. Substituting the dimensions of the variables and the appropriate indices into equation (11.6):

$$(ML^{-1}T^{-2}) = B[(L^a)(M^b L^{-3b})(M^c L^{-c} T^{-c})(L^d T^{-d})(L^e)].$$

The indicial equations are:

for M $1 = b + c$
for L $-1 = a - 3b - c + e + d$
for T $-2 = -c - d$

from which $b = 1 - c$
and $d = 2 - c$

$$\begin{aligned}
\text{now } a &= 3(1-c) + c - e - d - 1 \\
&= 3 - 3c + c - e - d - 1 \\
&= 3 - 2c - e - (2-c) - 1 \\
&= 3 - 2c - e - 2 + c - 1
\end{aligned}$$

and finally $a = -c - e$.

Substituting the indicial equations for a, b and d into equation (11.6):

$$\tau = B[(d^{(-c-e)}), (\rho^{(1-c)}), (\mu^c), (u^{(2-c)}), (k_s^e)].$$

There are three indices here: numerical, c and e and therefore three groups of variables

thus $\tau = B[(u^2 \rho), (\rho u d/\mu)^{-c}, (k_s/d)^e]$
and $\tau = \phi[(u^2 \rho), (\rho u d/\mu), (k_s/d)]$.

From this expression you will recognise the middle group as the Reynolds number and the last group as relative roughness (see section 6.5 of Chapter 6 for the introduction of this last term), both of which are dimensionless. The formula for shear stress can therefore be written as:

$$\tau = (u^2\rho), \phi[(\rho u d/\mu), (k_s/d)]$$

Solution (b)
Since the second and third groups are dimensionless the remaining dimensionless group will be: $(\tau/u^2\rho)$.

Buckingham's Pi theorem of $(n - m) = (6 - 3) = 3$ and therefore agrees with this solution. You should now confirm that the three groups of terms are dimensionless.

Example 11.5
The heat transfer by forced convection from a fluid transported in a long straight tube is governed by the variables $h, d, \mu, \rho, k, C,$ and u such that:

$h = \phi(d, \mu, \rho, k, C, u)$. Using dimensional analysis determine the form of the equation and the dimensionless grouping

Solution
If an exponential relationship is assumed in which B is the constant of proportionality and the indices a, b, c, e, f and g are numerical constants,

$$h = B[(d^a)(\mu^b)(\rho^c)(k^e)(C^f)(u^g)] \tag{11.7}$$

Introducing dimensions to each of the terms using Table 10.1 and including the indices as appropriate:

$$(MT^{-3}\theta^{-1}) = B[(L^a), (M^b L^{-b} T^{-b}), (M^c L^{-3c}), (M^e L^e T^{-3e} \theta^{-e}),$$
$$(L^{2f} T^{2f} \theta^{-f}), (L^g T^{-g})]$$

Collecting the indices:

for M	$1 = b + c + e$	(11.7a)
for L	$0 = a - b - 3c + e + 2f + g$	(11.7b)
for T	$-3 = -b - 3e - 2f - g$	(11.7c)
for θ	$-1 = -e - f$	(11.7d)

There are six unknown indices and four indicial equations. Evaluating in terms of the unknown indices c and f.

from equation (11.7d) $e = 1 - f$ (11.7e)

substitute equation (11.7e) into equation (11.7a)

$$1 = b + c + 1 - f$$

from which $b = f - c$ \hfill (11.7f)

Substitute equations (11.7e) and (11.7f) into equation (11.7c)

$$-3 = -(f - c) - 3(1 - f) - 2f - g$$
$$-3 = -f + c - 3 + 3f - 2f - g$$

from which $g = c$ \hfill (11.7g)

Substitute equations (11.7e), (11.7f) and (11.7g) into equation (11.7b)

$$0 = a - (f - c) - 3c + (1 - f) + 2f + c$$
$$0 = a - f + c - 3c + 1 - f + 2f + c$$

from which $a = c - 1$

Substituting the indicial equations for a, b, g and e into equation (11.7)

$$h = B[(d^{(c-1)}), (\mu^{(f-c)}), (\rho^c), (k^{(1-f)}), (C^f), (u^c)].$$

The variables are now related to the unknown indices c and f and index 1.0. There are therefore three groups of variables.

Thus $h = B[(k/d), (d\rho u/\mu)^c, (\mu C/k)^f]$.

Adopting Buckingham's Pi theorem there are $(n - m) = (7 - 4) = 3$ dimensionless groups here, thus by rearranging the equation:

$$(hd/k) = B[(d\rho u/\mu)^c, (\mu C/k)^f].$$

The group (hd/k) is known as the Nusselt number Nu;

the group $(d\rho u/\mu)$ is the Reynolds number Re;

and the group $(\mu C/k)$ is known as the Prandtl number Pr.

The value of the numerical constants B, c and f are found empirically (by experiment). It has been established that the values of B, c and f are constant for a very wide range of Re and Pr numbers and $B = 0.023$, $c = 0.8$ and $f = 0.33$.

The heat transfer correlation for flow of a fluid through a long tube is therefore:

$$Nu = 0.023(Re)^{0.8}(Pr)^{0.33}.$$

This formula is introduced in Chapter 3 as equation (3.10).

You should now confirm that the Nusselt, Reynolds and Prandtl numbers are dimensionless.

Example 11.6
Heat transfer by free convection in turbulent flow over vertical plates is governed by the variables $h, dt, \beta, g, d, \rho, \mu, k$, and C such that:

$h = \phi(dt, \beta, g, d, \rho, \mu, k, C)$. Using dimensional analysis determine the form of the equation and the dimensionless groups.

Solution
If an exponential relationship is assumed in which B is the constant of proportionality and the indices a, b, e, f, j and n are numerical constants,

$$h = B[((dt\beta g)^a), (d^b), (\rho^e), (\mu^f), (k^j), (C^n)]. \quad (11.8)$$

Introducing dimensions to each of the terms using Table 10.1, with the exception of the terms dt, β and g which are combined as shown in equation (11.8) to ensure that the unknown indices are limited to six, and including the indices as appropriate:

$$MT^{-3}\theta^{-1} = B[(L^aT^{-2a}), (L^b), (M^eL^{-3e}), (M^fL^{-f}T^{-f}),$$
$$(M^jL^jT^{-3j}\theta^{-j}), (L^{2n}T^{-2n}\theta^{-n})].$$

Collecting the indices

for M	$1 = e + f + j$	(11.8a)
for L	$0 = a + b - 3e - f + j + 2n$	(11.8b)
for T	$-3 = -2a - f - 3j - 2n$	(11.8c)
for θ	$-1 = -j - n$	(11.8d)

There are six unknown indices and four indicial equations. Evaluating in terms of the unknown indices a and n.

from equation (11.8a) $f = 1 - e - j$ (11.8e)
from equation (11.8d) $j = 1 - n$ (11.8f)

substitute equations (11.8e) and (11.8f) into equation (11.8c)
$\quad -3 = -2a - (n - e) - 3(1 - n) - 2n$
thus $-3 = -2a - n + e - 3 + 3n - 2n$
and $e = 2a$ (11.8g)

substitute equations (11.8f) and (11.8e) into equation (11.8b)
$\quad 0 = a + b - 3e - (1 - e - j) + (1 - n) + 2n$
thus $0 = a + b - 3e - 1 + e + j + 1 - n + 2n$
substitute equation (11.8f) $0 = a + b - 2e + (1 - n) + n$
substitute equation (11.8g) $0 = a + b - 4a + 1$
from which $b = 3a - 1$

substitute equations (11.8f) and (11.8g) into equation (11.8e)

$f = 1 - 2a - (1 - n)$

from which $f = 1 - 2a - 1 + n$

and therefore $f = n - 2a$

Substituting the indicial equations for b, e, f and j into equation (11.8)

$$h = B[((dt\beta g)^a), (d^{(3a-1)}), (\rho^{2a}), (\mu^{(n-2a)}), (k^{(1-n)}), (C^n)].$$

The variables are now related to the unknown indices a and n and numerical index 1.0. There are therefore three groups of variables.

thus: $h = B[(k/d), (dt\beta g d^3 \rho^2/\mu^2)^a, (\mu C/k)^n].$

Adopting Buckingham's Pi theorem there are $(n - m) = (7 - 4) = 3$ dimensionless groups here,

thus: $(hd/k) = B[(dt\beta g d^3 \rho^2/\mu^2)^a, (\mu C/k)^n],$

where

(hd/k) is the Nusselt number Nu,

$(dt\beta g d^3 \rho^2/\mu^2)$ is the Grashof number Gr,

and

$(\mu C/k)$ is the Prandtl number Pr.

The value of the numerical constants B, a and n are found empirically.

From Chapter 3, equation (3.6) for air uses the constants $B = 0.13, a = 0.33$ and $n = 0.33$ when $Gr > 10^9$,

thus $Nu = 0.1[(Pr)(Gr)]^{0.33}$.

You will notice that the number of variables n in Buckingham's Pi theorem assumes that the variables dt, β and g are kept together as one, as they are so designated at the commencement of this solution. You should now confirm that the Nusselt, Prandtl and Grashof numbers are dimensionless.

Summary for Examples 11.5 and 11.6
In both of these solutions you can see that the formulae for forced convection inside long tubes and free turbulent convection over vertical plates only require experimental work to evaluate the numerical constants. Dimensional analysis therefore can provide a fast track methodology for deriving empirical formulae in which the variables have been identified.

The following example relates to heat flow into a wall.

Example 11.7

A cavity wall constructed from two leaves of brick is subjected to a heat flux of I W/m² for time T seconds. The temperature rise $d\theta$ at the inside surface depends upon the physical properties of the lining which are its thermal conductivity k, density ρ and specific heat capacity C. The formula is likely to be in the form $d\theta = \phi[(I, T, k, \rho, C]$.

(a) By employing dimensional analysis find the form of the equation which will determine the rise in surface temperature of the wall after time T seconds.
(b) By adopting Buckingham's Pi theorem establish the dimensionless groups in the formula.
(c) Given that the numerical constant of proportionality B is $(2/\pi^{0.5})$, determine how long it will take to raise the inner surface temperature of the wall by 4 K. The width of the inner brick leaf is 100 mm and the temperature drop between its inner and outer surface is 3 K.

Take the thermal conductivity of the inner brick leaf of the wall as 0.62 W/mK, density 1700 kg/m³ and specific heat capacity 800 J/kgK.

Solution (a)
If an exponential relationship is assumed where B is the constant of proportionality and the indices a, b, c, d and e are numerical constants,

$$d\theta = B[(I^a), (T^b), (k^c), (\rho^d), (C^e)] \tag{11.9}$$

The dimensions of the variables in the proposed formula can be obtained from Table 10.1. In this solution, however, the additional dimension H for heat energy in joules will be used. Thus heat flux in J/sm² will have the dimensions $(HT^{-1}L^{-2})$.

Likewise thermal conductivity in J/smK will have the dimensions $(HT^{-1}L^{-1}\theta^{-1})$, and specific heat capacity in J/kgK will have the dimensions $(HM^{-1}T^{-1})$.

There will now be five dimensions used in this solution, namely H, M, L, T, θ. Introducing dimensions to each of the terms in equation (11.9) and including the indices where appropriate:

$$d\theta = B[(H^a T^{-a} L^{-2a}), (T^b), (H^c T^{-c} L^{-c} \theta^{-c}), (M^d L^{-3d}), (H^e M^{-e} \theta^{-e})].$$

Collecting the indices:

for H, $0 = a + c + e$ (11.9a)
for M, $0 = d - e$ (11.9b)
for L, $0 = -2a - c - 3d$ (11.9c)
for T, $0 = -a + b - c$ (11.9d)

for θ, $1 = -c - e$ (11.9e)
from equation (11.9b) $d = e$ (11.9f)
from equation (11.9d) $a = b - c$ (11.9g)
From equation (11.9e) $e = -1 - c$; substitute this and equations (11.9g) into equation (11.9a)
$$0 = b - c + c - 1 - c$$
from which $1 = b - c$ (11.9h)
substitute equation (11.9h) into equation (11.9d): $0 = -a + 1$
from which $a = 1$ (11.9i)
Add equations (11.9a) and (11.9c): $0 = -a - 3d + e$
substitute equation (11.9f): $0 = -a - 3e + e$
$$0 = -a - 2e$$
substitute equation (11.9i): $0 = -1 - 2e$
from which $e = -0.5 = d$ (11.9j)
substitute equation (11.9j) in equation (11.9e): $1 = -c - (-0.5)$
from which $c = -0.5$ (11.9k)
substitute equations (11.9i) and (11.9k) into equation (11.9d):
$$0 = -1 + b - (-0.5)$$
from which $b = 0.5$ (11.9l)

The indices a, b, c, d and e each have numerical values as identified in equations (11.9i), (11.9l), (11.9k) and (11.9j).

Substituting these indicial values into equation (11.9):

$$d\theta = B(IT^{0.5}k^{-0.5}\rho^{-0.5}C^{-0.5})$$

thus $d\theta = B[(I)(T/k\rho C)^{0.5}]$.

In this solution the only numerical value which must be found empirically is the constant B since the numerical values of the indices a, b, c, d and e have been evaluated during the process of analysis.

Solution (b)
From Buckingham's Pi theorem $(n - m) = (6 - 5) = 1$ dimensionless group. Rearranging the formula derived in part (a):
$$1 = B[(I/d\theta)(T/k\rho C)^{0.5}]$$ (11.10)

The dimensions of the terms are now analysed:
the dimensions of the terms $(I/d\theta)$ are: $(HT^{-1}L^{-2}\theta^{-1})$,
the dimensions of the terms $(T/k\rho C)^{0.5}$ are:

$$(TH^{-1}TL\theta M^{-1}L^3 H^{-1}M\theta)^{0.5},$$

this reduces to:

$$(T^2 H^{-2} L^4 \theta^2)^{0.5}$$

accounting for the index of 0.5, the dimensions of the terms $(T/k\rho C)^{0.5}$ are:

$(TH^{-1}L^2\theta)$.

Now combining the dimensions of the terms: $(I/d\theta)(T/k\rho C)^{0.5}$ we have:

$(HT^{-1}L^{-2}\theta^{-1}TH^{-1}L^2\theta)$ from which this group is dimensionless.

Summary of parts (a) and (b) of Example 11.7
You can see by a process of cancellation, the combined group of terms in equation (11.10) is dimensionless and Buckingham's Pi theorem confirms this. The numerical constant B is also dimensionless. The index of 0.5 in the formula is accounted for in the determination of the dimensionless group since it was evaluated during the process of deriving the equation. In the case of Examples 11.5 and 11.6 the indices have to be found empirically and therefore do not form part of the analysis of the dimensionless groups.

Solution (c)
Adopting the formula obtained in part (a): $d\theta = (2/\pi^{0.5})(I)(T/k\rho C)^{0.5}$. Rearranging the equation in terms of time T in seconds by first of all squaring both sides:

$(d\theta)^2 = (4/\pi)(I)^2(T/k\rho C)$.

thus $T = [(d\theta)^2 \pi k\rho C]/(4I^2)$.

Heat flux I, from Fourier's equation (section 2.2, Chapter 2) will be:

$I = k dt/d = (0.62 \times 3)/0.1 = 18.6 \, \text{W/m}^2$.

Substituting: $T = (4^2\pi \times 0.62 \times 1700 \times 800)/(4 \times 18.6^2)$

from which $T = 30\,628\,\text{s}$

and the time taken to raise the inner surface temperature of the inner brick leaf of the wall by 4K will be 8 h 30 min.

11.5 Chapter closure

Successful completion of this chapter provides you with underpinning knowledge in respect of some of the formulae employed in fundamental calculations relating to the design processes in heat and mass transfer in the subjects of heating, ventilating and air conditioning. Dimensional analysis is not a wonder tool. Perhaps you will have noticed in the examples selected that a knowledge of the processes to be analysed is essential when it comes to finding the form of an equation relating to a number of variables.

Bibliography

Aubrey and Burstall, *A History of Mechanical Engineering*, Faber, 1963.

H. B. Awbi, *Ventilation of Buildings*, E&FN Spon, 1995.

Building Services, the CIBSE Journals: July, 1992, Thermal comfort compliance. June, 1993, Thermal comfort; UK design rules OK?. January, 1994, A bouquet of barbed wire. November, 1994, De Montfort University School of Built Environment. December, 1994, Advancing the cause of low energy design. February, 1995, Which ventilation system. April, 1995, East Anglia University new academic building. May, 1995, BRE low energy office. June, 1995, Control naturally. BSRIA research project. July, 1995, Navigating the comfort zone. November, 1995, No. 1 Leeds City Office Park. January, 1996, Canterbury Court Centre. February, 1996, Ventilation solutions. March, 1996, Future buildings. April, 1996, Green buildings; benefits & barriers. June, 1996, Cable and Wireless College. Probe 5. August, 1996, Probe 6, Woodhouse Medical Centre. November, 1996, Refurbishment the natural way. November, 1996, Notes on European ventilation rates for buildings. December, 1996, Probe 8, Queen's Building, Anglia Polytechnic University. January, 1997, Salt-bath modelling of air flows. February, 1997, Portsmouth University, Portland Building. March, 1997, Green Demo, environmental office of the future.

D. V. Chadderton, *Air Conditioning, a Practical Introduction*, E&FN Spon, second edition, 1997.

Chartered Institution of Building Services Engineers, *CIBSE Guide*, Books A and C.

J. F. Douglas, *Solutions to Problems in Fluid Mechanics*, Part 1, 1975; Part 2, 1977. Pitman

O. Fanger, *Thermal Comfort*, Macgraw Hill, 1972.

J. A. Fox, *An Introduction to Engineering Fluid Mechanics*, Macmillan, 1977.

Heating and Air Conditioning Journal. Eight articles on Industrial Energy Efficiency: part 1, The scope for energy saving, July 1982; part 2, Radiant heaters and heat recuperators, September 1982; part 3, Liquid/liquid heat exchangers, October 1982; part 4, Plate and spiral flow heat exchangers, November 1982; part 5, Heat exchangers for gases, December 1982; part 6, Thermal wheels, January 1983; part 7, Assessment of industrial techniques, February 1983; part 8, Heat pipes, March 1983.

Institute of Plumbers, *The Plumbing Guide*.

Memmler, Cohen and Wood, *The Human Body in Health and Disease*, Lippincott, 1992.

K. J. Moss, *Heating and Water Services Design in Buildings*, E&FN Spon, 1996.

Rogers and Mayhew, *Thermodynamic and Transport Properties of Fluids*, Blackwell, 1995

J. R. Simonson, *Engineering Heat Transfer*, Macmillan, 1981.

Index

Page numbers in **bold** indicate figures, and numbers in *italic* tables

Absolute roughness 118, *148*
Absorptivity 68
Air flow through openings 171
Air heater battery 201
Aluminium foil, use of 79, 81, 85
Air movement around buildings **166, 167, 168**
Air pressure distribution resulting from wind **169**
Air temperature 10

Bellmouth 138, 150, 151
Bernoulli equation 96, 120, 130, 132, 133, 135, 136, 139, 140, 141, 146, 147, 150, 152, 153, 158, 170
Black body emissive power **72**
Black body radiation 68
Body heat loss 4, **5**, 8
Boiling 189
Boosted water 145
Boundary layer formation **115, 116**
Boundary layer separation 117
Boundary surface 144, 185, 195
Box's formula 113, 125, 210
Bridges, heat (discrete, multiwebbed, finned) 37, 39, 40, 41
Buckingham's theorem 215, 220

Calandria 195
Calibrating a differential manometer 98
Calibrating a manometer 97
Capacity ratio 196
Cavity wall 226
Celsius scale 3
Chezy coefficient 155, 156, 157

Chezy formula 156, 157, 159, 160, 161, 162, 163
Clo, unit of *15*
Coefficient of friction 113, 118, 119, **120**, 121, 126, 150, 162
Coefficients of friction, comparison 162
Colour temperature guide *71*
Comfort temperature (dry resultant temperature) 11
Colebrook–White Moody chart **120**
Comfort diagrams **15, 16**
Comfort scales 19
Comfort zones 14
Condensing 190, 199
Conductive heat flow through a wall 30
Conductive heat flow variations 30
Cooling by ventilation 179
Colebrooke–White formula 119
Continuity of flow 95
Contraction, sudden **138**, 140
Counterflow 187, 188, 191, 194, 199, 202, 205
Crimp and Bruge's formula 162, 163, 164
Cross ventilation
  stack effect 175, 176, **177**, 178, 179
  wind effect 171, 172, **176**, 177, 179
Cubicle expansion of air 52

Darcy formula 110, 121, 127, 144, 148, 158, 213
Darcy–Weisback formula 162, 163, 164

Density of materials 26
Differential manometer 98, 139, 140, 142
Dimensions of derived units **209**
Double pipe heat exchanger 191
Drain discharge capacity 156
Drain gradient 157, 160
Dynamical similarity 216, 220

Eccrine glands 7
Economizer 197
Edge loss from ground floors 33
Effectiveness 196, 197, 199, 201
Electromagnetic radiation **69**, 71
Emissivity 68, **70**, 74, 75, 77, 79, 81, 84, 85
Energy 2
Environmental temperature 11

Face and interface temperatures 27
Fan laws 217
Fan power 219
Filter replacement 98
Fire hydrant 133, 136
Flow in vertical soil stacks 158, 159
Flue gas temperature measurement 87
Fluid velocity comparisons 163, 164
Foot pound second (FPS) system 208
Forced convection inside straight tubes **51**, 53, 60, 64, 222
Form factor 75, 79, 81, 85, 87

Fouling factors 191
Fourier's law 27, 44
Free convection over vertical plates 51, 53, 57, 58, 62, 63, 224

Gate valve, pressure loss 142
Geometric similarity 215, 220
Globe temperature **11**
Globe valve pressure loss 141
Grashof number 52, 53, 54, 225
Gravitational flow, pipes in parallel 151, 154
Gravitational flow of water 150, 153
Gravitational mass transfer 157, 158, 160
Grey surface 72
Ground loss 33

Hagen 110
Heat absorption into a wall 226
Heat exchangers **185**
Heat flow, hollow blocks 39, 41
Heat flow paths 28, 29, 33, 35
Heat flux 20, 63, **73**, **77**, **78**, **79**, 84, 85, 87,
Heat loss from a thermal storage vessel 43
Heat loss from a pipe carrying steam 45
Heat radiation exchange **76**, 79, 81, 84, 85, 87
Heat radiation, spectral proportions 72
Heat radiation waveform **69, 71**
Heat transfer 2
Heat transfer coefficient 29, 30, 31, 53, 57, 58, 60, 62, 63, 64, 76, 79, 81, 84, 87, 222, 224
Heat transfer inside straight tubes 53, 222
Hydraulic gradient 143, 155, 156, 157, 158, 161
Hydraulic loss 143, 149
Hydraulic mean diameter 156
Hyperpyrexia 6
Hypothalamus 7
Hypothermia 5

Impermeability 159
Incident heat radiation on a surface 70, 89, 91, 92
Inclined differential manometer 99
Inclined venturi meter **105**

J values for windows *177*

Kata thermometer 11
Kelvin scale 3
Kinetic energy 96, 130
Kirchoff's law 74

Lambert's law 74, **77**
Laminar flow 110, **111, 112,** 116, 121, 122, 126, 128
Laminar sublayer **115, 119**, 195
Latent heat loss **6**, 8
Line source heat radiation 77
Log mean temperature difference 54, 182, 186, 188, 191, 194, 201, 202, 205
Loss, hydraulic, in straight pipes, fittings, etc. 137, 138, 139, 140, 141, 143, 145, 147, 149, 150, 151, 153
Luminous quartz heater emission 78

Manning formula 162, 163, 164
Manometer **97**
Mean bulk temperature 54
Mean film temperature 54
Mean radiant temperature 10
Mean temperature difference 54
Met 4, 5, 15
Metabolic rate 4, 5
Metre kilogram second (MKS) system 208
Minimum thickness of insulation 46, 47
Monochromatic emissive power 73

Natural draught **173**
Natural ventilation, influencing factors 166

Natural ventilation of internal space subject to temperature difference 177
Natural ventilation of internal space subject to wind 176
Non-standard U value for bridged wall 38
Number of transfer units 196
Nusselt number 52, 53, 54, 57, 58, 60, 64, 223, 225

Orifice plate **101**, 103
Osborne Reynolds' experiment **111**
Overall U value (thermal transmittance) 56, 60, 64, 192, 194, 199, 201, 202, 205

Panel radiator, convective emission 57
Parallel flow **186**, 191, 192, 197
Pipe coil, emission from 58
Pipe sizing, rising main 124
Pitot static tube **106**
Pitot static tube, constant 107
Planck's law 74
Plane radiant temperature 89
Point source heat radiation 77
Poiseuille's equation 110, 122, 124
Potential energy 96, 130
Prandtl number 52, 53, 54, 56, 58, 60, 65, 115, 223, 225
Predicted mean vote 19
Predicted mean vote index **21**
Predicted percentage dissatisfied 19
Pressure energy 96, 130
Pressure loss along a pipe 121
Pressure loss in a system 143
Pump duty, net 145
Pump laws 217
Pumping power 122, 145, 219

Radiant heat, efficiency 87
Radiant panel emission 84
Radiation heat flux ratio 72
Radiative emission from black, grey and selective surfaces 73

Index

Rainfall run-off 160
Rate of cooling by ventilation 179
Reflectivity 68
Refrigeration condenser 194, 205
Relative roughness 118, **119**, 120, 148
Reynolds experiment **111**
Reynolds number 52, 53, 54, 110, 113, 119, 121, 122, 126, 128, 144, 148, 149, 210, 223
Rising main, sizing 124, 147
Roughness coefficient 162
Roughness ratio, relative roughness **120**, 149, 150

Scale model (fan) 220
Selective surface 72
Sensible heat loss 6, **8**
Shear stress 221
Shell and tube heat exchanger 192
Shock loss 137, 138, 139
Sludge flow 216
Soil stack sizing 158
Solar collector efficiency 91
Solar collector energy absorption 92
Solar constant 90
Solar intensity 90

Specific heat capacity of materials **26**
Steady flow 95
Stefan–Boltzman constant 73, 211
Stefan–Boltzman law 74
Static pressure loss along a duct 126, 149
Static regain 134, 144
Storage calorifier 199
Streamlines in laminar flow **111**
Subcutaneous tissue 9
Suction lift 130
Surface characteristics *70*
Syphon 132
Système Internationale 208

Temperature 2, **3**
Temperature gradient through a wall 31
Thermal conductivity 25, **26**
Thermal index 9
Thermal lag 228
Thermal resistance 27
Thermal transmittance coefficient 27
for floors 34
Thermometer types 3, 4
Thermometry 3
Thickness of duct insulation 47
Thickness of pipe insulation 46
Total energy 211

Transmissivity 68
Transverse wave motion 69
Tube bundle 205
Turbulent flow **112**, 114, 121, 124, 126, 143, 149

Uniform flow 95

Vasoconstriction 9
Vasodilation 9
Vector radiant temperature 89
Velocity gradient **114**
Velocity head loss factor 139, 140
Velocity of wave propagation 74
Velocity pressure loss factor 138, 142
Velocity profile **112, 116**
Ventilation rate 172, 175, 178
Venturi meter 100, 102, 103, 105
Viscosity 109
Volume flow rate of air in a duct 108

Wein's displacement law 73, 74
Wet bulb temperature 10
Wind flow patterns **166, 167, 168, 169, 170**
Wind loading pattern on a building facade **169**